W9-ADK-148

# The History, Psychology, and Pedagogy of Geographic Literacy

Malcolm P. Douglass

Westport, Connecticut
London

**Library of Congress Cataloging-in-Publication Data**

Douglass, Malcolm P. (Malcolm Paul), 1923–
The history, psychology, and pedagogy of geographic literacy / Malcolm P. Douglass.
p. cm.
Includes bibliographical references and index.
ISBN 0–275–96138–9 (alk. paper)
1. Geography—Study and teaching. 2. Geography. I. Title.
G73.D68 1998
910'.7—dc21 98–14489

British Library Cataloguing in Publication Data is available.



Library of Congress Catalog Card Number: 98–14489
ISBN: 0–275–96138–9

First published in 1998

Praeger Publishers, 88 Post Road West, Westport, CT 06881
An imprint of Greenwood Publishing Group, Inc.

Printed in the United States of America

The paper used in this book complies with the Permanent Paper Standard issued by the National Information Standards Organization (Z39.48–1984).

10 9 8 7 6 5 4 3 2 1

**Copyright Acknowledgments**

The author and publisher are grateful to the following for permission to reproduce copyright material:

Beacon Press and Victoria Shoemaker Agency, from *The Geography of Childhood* by Gary Paul Nabhan and Stephen Trimble (Boston: Beacon Press, 1994). Copyright © 1994 by Gary Paul Nabhan and Stephen Trimble. Used by permission of Beacon Press, Boston, and Victoria Shoemaker Agency.

Random House, from Charles Pellgrino, *Unearthing Atlantis: An Archaeological Odyssey* (New York: Random House, 1991). Copyright 1991 by Charles Pellegrino.

The National Geographic Society, for reproducing findings from the 1988 Gallup Survey in *Geography: An International Survey* (Washington, DC: National Geographic Society, 1988).

National Society for the Study of Education, for Figure 1.2, from Harold Rugg, ''Three Decades of Mental Discipline: Curriculum Making via National Committees,'' in Guy Whipple, ed., *Curriculum Making: Past and Present*, Twenty-sixth Yearbook of the National Society for the Study of Education, Part 1 (Bloomington, IL: Public School Publishing Company, 1926), pp. 48–49.

Houghton Mifflin Company, for Figure 1.3, from Paul R. Hanna, Rose E. Sabaroff, Gordon F. Davies, and Charles R. Farrar, *Geography in the Teaching of Social Studies: Concepts and Skills* (Boston: Houghton Mifflin Company, 1966). Copyright © 1966 by Houghton Mifflin Company. Used with permission.

Hoover Institution Press, for Figure 1.4, reprinted from *Assuring Quality for the Social Studies in Our Schools*, by Paul Robert Hanna (Stanford, CA: Hoover Institution Press, 1987), with the permission of the publisher, Hoover Institution Press. Copyright © 1987 by the Board of Trustees of the Leland Stanford Junior University.

Little, Brown and Company, for Figure 2.1, from *The Story of Maps* by Lloyd A. Brown (Boston: Little, Brown and Company, 1949). Copyright © 1949 by Lloyd Arnold Brown; © renewed 1977 by Florence D. Brown. By permission of Little, Brown and Company.

Topkapi Palace Museum, for Figure 2.4. Reproduced with permission.

Barnes & Noble, for Figure 2.6, from Arild Holt-Jensen, *Geography: History and Concepts*, 2nd ed. (Totowa, NJ: Barnes & Noble, 1988). Reproduced with permission.

The British Museum, for Figure 3.1. Reproduced with permission.

Routledge, for Figures 3.2 and 3.3, from Jean Piaget and Barbell Inhelder, *The Child's Conception of Space* (London: Routledge and Kegan Paul, 1956). Reproduced with permission.

Department of Geography, Western Michigan University, for Figure 3.4, from Joseph Stoltman, ed., *International Research in Geographic Education: Spatial Stages of Development in Children and Teacher Classroom Style in Geography* (Kalamazoo: Western Michigan University Department of Geography, 1976). Reproduced with permission.

Routledge, for Figures 3.5 and 3.6, from Peter Gould and Rodney White, *Mental Maps*, 2nd ed. (Boston: Allen & Unwin, 1986). Reproduced with permission.

The University of Chicago Press, for Figures 4.1 and 4.3, from David Greenhood, *Mapping* (Chicago: University of Chicago Press, 1964). Reproduced with permission.

Universal Press Pty. Ltd., for Figure 4.2. Reproduced with permission of UP Pty. Ltd. Copyright Universal Press Pty. Ltd., 1/98 CEN.

Harveys, for Table 5.1, from Carol McNeill, Jean Ramsden, and Tom Renfrew, *Teaching Orienteering: A Handbook for Teachers, Instructors, and Coaches* (Perthshire, UK: Harveys, 1987).

Every reasonable effort has been made to trace the owners of copyright materials in this book, but in some instances this has proven impossible. The author and publisher will be glad to receive information leading to more complete acknowledgments in subsequent printings of the book, and in the meantime extend their apologies for any omissions.

*This book is dedicated to my children,*
*Susan Enid, John Aubrey, and Malcolm Paul.*
*I admire you, I learn from you, and I love you.*

# Contents

# Figures and Tables

## FIGURES

## TABLES

# Preface

The idea of geography as particularly worthy of a place in the school curriculum is experiencing a kind of rebirth. It is not that this "mother of all the sciences" has ever been totally excluded as an educational good, as something unworthy or not important enough to claim its share of the instructional day, but it has experienced a checkered career as a subject in the school curriculum. Now, however, under the auspices and vast wealth of the National Geographic Society, and kindred organizations that share the society's goals, there is widespread interest in the problem of geographic literacy and in developing strategies for accomplishing the goal of bringing into being a significantly more literate geographic society.

The most pervasive problem facing those who would, in a sense, resurrect geography has two aspects. The first of these has to do with the problem of understanding what actually constitutes this thing called *geography* (*geo*/"earth,"/*graphy*/"writing"). The public generally, and the teaching professions as well, tend to think of geography, at least as far as our elementary and secondary schools are concerned, as the compilation of facts. Particularly, this means the ability to name and locate phenomena, usually on a map, but generally it implies the ability to answer what I have called "the Mozambique question"; that is, the "Where is it and what is it?" kind of question. The fact that students in our schools, and the public at large, are, more likely than not, themselves unable to answer the Mozambique question has been offered up as proof of the illiteracy that has plagued Americans from time immemorial, or at least since such questions were first asked in any formal sort of way. (Chapter 1 describes the first such survey, which was conducted in the 1840s in the Boston Public Schools.) But is this a true indicator of geographic knowledge—of geographic literacy?

I have tried in the pages that follow to disabuse the reader of this view. Geographic literacy involves knowledge of facts and even bits of specific information, of course, but it is a much larger matter than the simple acquisition of facts; it is, rather, a way of knowing, a process in which literacy evolves as one learns to think in terms of spatial interactions. Facts, other specific bits of information, and skills are necessary to undergirding geographic literacy, but they become meaningful, and persist in one's memory, only when they participate in the development of a larger whole.

Teachers, parents, school administrators, and the public are, as a consequence, generally and specifically handicapped by a point of view—even a prejudice—that restricts thinking about what being geographically literate entails, and therefore discourages detailing the parameters of ways that might lead to its achievement. Thus, the second aspect of the problem of geographic literacy is to deal with the question of what is important to know that may lead us beyond this very narrow notion of what has come to be called "school geography" (in contrast with what real geographers know and do in the practice of their profession).

The task is a large one for a variety of reasons, the most pervasive of which is that it is impossible to teach effectively without a sound knowledge of what it is one wants to teach. Thus, the problem concerns, not just teacher education itself, but moving the minds of those who supervise teachers, those who plan curriculums, and those many others who make decisions that affect what teachers do in the classroom, including publishers of textbooks. For a variety of reasons which I explore here, the profession of geography is relatively small, fractionated, and, particularly unfortunate for the teaching profession, dominated by practitioners who are overwhelmingly male.* It is consequently not in a position to communicate with a broad enough band of teachers and persons in teacher preparation programs to provide a viable or empathic cadre of informed professionals.

This book is, therefore, an attempt to (at least partially) bridge the gap between geographers and those who hold responsibilities for teaching our children and youth. Its purpose is to provide a content base for decision making; it is not a text or a how-to book regarding the teaching of geography. Generally, however, I hold to the view that the separate subject curriculum, which if followed would instate "geography" as a separate course of study, is not a viable answer, either where the "subject" is geography or, speaking in more general terms, where the curriculum as a whole is concerned. But if there is a particularity to geography that makes the case stronger for some form of integration,

*Rose Gillian's book, *Feminism and Geography: The Limits of Geographical Knowledge* (Minneapolis: University of Minnesota Press, 1993), p. 1, reports a profile of the Association of American Geographers showing that only 18.6 percent of the membership who were employed by colleges and universities were women. In addition, of that number, a disproportionately large portion of that percentage held less-influential, temporary, part-time, and/or lower-paid positions within the departments.

it is that, as a way of thinking about phenomena, *geography* in any event permeates the curriculum as it presently stands. There are few, if any, subjects that cannot be considered from a geographical point of view. Geography, like history, is ultimately more a way of thinking about things than it is a body of specific information. The catch, of course, is that the teacher must be sufficiently informed to see the possibilities.

I have approached the problem with three themes in mind. One is historical. What seems cogent about the history of curriculum development, and the role of geography therein, so that a place for geographic inquiry can be found within it? The history of curriculum reform in the United States since the organizing of what we now might call the modern American school system (dating as it does from the period following the Civil War) has, for a variety of reasons, not been kind to the geographic strand of the school's curriculum. I explore what seem to me to be central aspects of this problem in Chapter 1. Then, I ask, what seems important about the evolution of geographic thought; what might be helpful to know about the origins of geographic thinking, its development over time, and how this contrasts with what geography is today? I have proceeded here on the assumption that the basic ideas of geography can be gleaned from the historical record and that a context for thinking about geography in the classroom gains richness with such knowledge. This constitutes the basic thrust of Chapter 2.

In the next two chapters, I turn to the question of what the psychology of learning may reveal to us about how spatial concepts are acquired and how they might be developed. In the first of these, Chapter 3, I bring together the research on the learning of spatial concepts. The great pioneer in this area is Jean Piaget, the Swiss epistemologist/psychologist. The publication of his initial studies into this aspect of learning (in France in 1948, but not in England until 1963 and in the United States even later) stimulated further research, but only for a limited time. Unhappily, little has been done to extend those findings in the last two decades, so there is doubtless much more to be learned in this regard. In Chapter 4 the reader will find a discussion of aspects of the learning-to-read process where map reading is concerned. I approach the problem of map reading as one that is similar to but more complex than print reading. It is essentially the process of creating meanings for the various representations appearing on maps, including not only those symbols that stand for physical and human characteristics in the landscape but also those subtle elements of distance, direction, and distortion that must be interpreted if meaningful reading is to take place.

The next section (Chapter 5) treats geography as an aspect of the total curriculum. The idea of geography is not limited to a ''subject'' that can be isolated from the rest of the curriculum. Geography is everywhere, and it has been my purpose in this portion of the book to illustrate this fact with examples from science and mathematics, the social studies, the arts, and even the world of sports. Chapter 6 deals with the problem of evaluation. The field of educational evaluation is in transition, as it is moving from a quantitative to a qualitative emphasis. Many problems have arisen as this transformation is being attempted, raising questions ranging from ways of conducting in-depth assessment while attempting to adhere to the American penchant to test everyone with considerable frequency to those involved in the problem of questioning that may be perceived as querrying too deeply into student values and motivations. Americans are particularly touchy on this latter point, providing a complicating element in developing tools that help us understand more fully the effectiveness of our teaching.

Finally (Chapter 7), I deal with some futuristic questions. What, for instance, will or should be the role of geographic teaching where cyberspace is concerned? What are some of the other concerns we need to take into account as we view geography as a unifying concern, a way of looking at the world, rather than as a set of specific bits of information and skills?

I have approached the problem of understanding the elements involved in the development of a geographically literate society from an interdisciplinary perspective. I do not believe any particular set of professionals—in this case, either geographers, psychologists, or educationists—has an exclusive hold on what is, or is not, important in approaching an educational issue such as this. It is a principle that I believe applies as well to the other traditional curriculum areas that continue to typify our elementary and secondary school curriculums, in the face of increasing acceptance of the fact that knowledge and knowing cannot be classified in this way. While I have argued that reality translates itself into a curriculum that is not integrative but, rather, separated into sets of discrete units, it remains true that most teachers teach the best they can under the circumstances in which they find themselves. Some of those circumstances have to do with the way a particular school is organized and its ethos, from the traditions that have evolved over time, the kind of leadership the ''front office'' provides, and, perhaps most central of all, the kind of colleagueship that characterizes the faculty itself. Perhaps even more important, however, is the stage of development in which the individual teacher finds him- or herself. Most teachers realize that any act of teaching is rarely fully satisfying and that, in any case, the process of becoming a teacher is a developmental one. Thus it is that in planning and teaching, each individual takes what he or she can from a variety of sources at any given moment. In that portion of the curriculum that lends itself to the development of students' understanding of spatial interactions, one can expect no more nor less, given the variables within which a teacher practices her or his profession.

# 1

# An Introduction to Geography in the School Curriculum

> What's the world coming to? A lot of Americans don't know where Mexico is. They think Delaware is a city. Asked to name a tribe that has invaded England, they answer, "The Aztecs." They ask Robert Young for medical advice, just because he played Dr. Marcus Welby on television, and they write letters to Kentucky Derby winners.
>
> "Not the jockey, not the trainer, but the horse," said an incredulous Steve Allen.
>
> With a look of inspired mystification, the veteran comedian-author-songwriter paused in his description [before an audience at Cal Tech], bolstered by studies and personal experiences, of creeping dumbness in America to picture somebody actually writing a letter to a horse. "Something like, 'Dear Seabiscuit. . . . Thanks for winning the Kentucky Derby. I won 28 bucks on you. Keep up the good work. . . .' "
>
> "The American people are dumber now than they have been in a very long time," said Allen, who has written a book called *Dumpth* on the subject.
>
> *Los Angeles Times*, January 21, 1990

Proving whether people are in fact "dumber" about geography, history (another area in which Allen claims Americans are suffering from "incipient amnesia"), or other domains in which knowledge is deemed to be critical is not as easy as one might think. What we do know is that the history of American education has continuously recorded expressions of dismay over the school's success in teaching what the adults of the community at any particular time have considered important for students to learn. And so has it been for geography, a school subject that once seemed in danger of fading from the curriculum forever, like

Latin, but which, like Phoenix, is now competing for attention after years of neglect amid choruses of complaint about the effectiveness of the school's efforts to educate for geographic literacy.

## THE STATE OF GEOGRAPHIC LITERACY

Americans are not people to sit by when they spot a "problem." We began early, barely 25 years after our Puritan forefathers set foot on (or near) Plymouth Rock. The issue then was dismay over the success, or lack thereof, of parents (particularly fathers) to teach their children (especially their sons) to read the Bible, a skill then universally considered essential to escaping the tortures of Hell. The result was passage in the legislature of the day of a 1647 bill called *Ye Olde Deluder Satan Act*, which provided for the establishment of tax-supported schools, a novel idea at the time.

In the next 200 years, American education branched far afield from its early concentration on matters religious. A vigorous, growing nation needed a more broadly educated citizenry. A national identity needed affirmation through history. A vibrant economy required a population with at least rudimentary skills in arithmetic, along with a knowledge of the world, particularly that world with which a burgeoning commercial trade was increasingly being conducted. And an expanding population complicated the growing commitment, as yet far from realized, to the idea of universal education, at least through the ages of fifteen or so.

It was this school world that Horace Mann (1796–1859) sought to improve, in 1845, in what turned out to be the first attempt to evaluate in a formal way the teaching effectiveness of Boston's public schools. In a test constructed to serve this end, Mann devised the first means for assessing what youngsters were learning as a consequence of formal instruction. While he and his Boston School Committee sought to evaluate learning in all aspects of the curriculum, we will focus our attention here on the kinds of geographic learning expected of eighth graders of that day. Not surprisingly, but to his horror, Mann met up with what he judged to be abysmal ignorance (a conclusion not limited to the geography curriculum). For example, to the question, "On which range of mountains is the line of perpetual snow most elevated above the ocean—on the Rocky Mountains of North America, or on the Cordilleras of Mexico?" Mann bemoaned the fact only 91 pupils could answer correctly (the Cordilleras), while 154 answered wrongly and another 242 did not even attempt an answer (quoted in Caldwell & Courtis, 1925, pp. 254–255). In another test item, the "scholars," as Mann and others referred to public school students of the time, were asked to draw an outline map of Italy. Mann wrote that, while "many attempted [to draw] it, . . . of the whole number [500], only seventeen made a drawing which could have been recognized as a representation of Italy by one who did not know what the scholar was trying to do" (Caldwell & Courtis, 1925, p. 259).

To the question, "Do the waters of Lake Erie run into Lake Ontario, or the waters of Ontario into Erie?'' 287 students answered correctly, but 130 answered incorrectly, and 72 failed to respond (Caldwell & Courtis, 1925, p. 255). Mann rightly commented, however, that this question was open to guessing; the ''scholars'' had as great a chance of getting the correct answer as the wrong one.

These are the other questions included on the test:

1. Name the principal lakes in North America.
2. Name the principal rivers in North America.
3. Name the rivers running eastward into the Mississippi.
4. Name the rivers running westward into the Mississippi.
5. Name the states which lie upon each bank of the Mississippi, and their capitals.
6. Which is most elevated above the level of the sea, Lake Superior or Lake Huron?
7. Write down the boundaries of Lake Erie.
8. Quebec is (according to your maps) 4° 40' north from Boston; Ithaca in New York, is 5° 30' west from Boston. Which place is farthest from Boston?
9. What is the general course of the rivers in North and South Carolina?
10. What is the general course of the rivers in Kentucky and Tennessee? What is the cause of the rivers in these four contiguous states running in opposite directions?
11. Which is most accessible in its interior parts, to ships and to commerce, Europe or Africa?
12. Name the empires of Europe.
13. Name the kingdoms of Europe.
14. Name the republics of Europe.
15. What is the nearest route from England to India—by Cape of Good Hope, or by the Red Sea?
16. The city of Mexico is in 20° of N. latitude; the city of New Orleans is in 30° of N. latitude. Which has the warmest climate?
17. Name the rivers, gulfs, oceans, seas and straits through which a vessel must pass in going from Pittsburgh in Pennsylvania, to Vienna in Austria.
18. On which bank of the Ohio is Cincinnati, on the right or left?
19. What are the principal natural and artificial productions of New England?
20. Over what continents and islands does the line of the equator pass?
21. What parts of the globe have the longest days?
22. If a merchant in Moscow dines at 3 o'clock, p.m., and a merchant in Boston at 2 o'clock, which dines first?
23. Name the countries which lie around the Mediterranean Sea.
24. What countries lie around the Black Sea?
25. What rivers flow into the Black Sea?

26. Name the principal ports of Russia on the Black Sea, on the White Sea, and on the Gulf of Finland. (Caldwell & Courtis, 1925, pp. 276–278)

Now let us move forward, over a period of almost 150 years to the recent past, when claims similar to Mann's were still being voiced. It is useless today to chronicle them all; they have been numerous. Let it only be said that, like all things regarding the educational process, they have come in waves or cycles. Often, their reappearance has been related to international events and to national pride and, perhaps, concern over the apparent ignorance of the American citizenry of places and events within and beyond the national borders.

The history of education has frequently been characterized by complaints of the ineffectiveness of the schools' attempts to inculcate information deemed, at least by the intellectual leadership of the country, to be important. No content area of the curriculum has escaped such criticism. Witness, for example, the furor over the assumed failure of the schools to prepare students in the sciences, which, while simmering for some time on the social stove, ignited with considerable heat when the USSR launched *Sputnik* in 1957.

The most recent large-scale international inquiry into the state of geographic education in American schools was completed in the late 1980s by the Gallup Organization, the well-known interviewing group, for the National Geographic Society (National Geographic Society, 1988; see also *The Geography Learning of High-School Seniors*, a 1990 report in the National Assessment of Educational Progress series of publications from the U.S. Office of Education which is discussed in this chapter). The intent of the survey was to gain information in two general areas: (1) the importance attributed to what was termed "knowledge of geography," and (2) "the extent of basic geographic literacy among adults 18 and over in the United States as well as in eight 'comparison countries' " (National Geographic Society, 1988, p. 2).

In evaluating the extent to which "Americans have an awareness of the importance of geography, its usefulness, and its influence on a variety of events and activities," it was found that:

> Nine in ten (90%) American adults think it is important to know something about geography in order to be considered a well-rounded person.
>
> Seven in ten (69%) believe it absolutely necessary to be able to read a map; significantly more than those who feel the same way about being able to write a business letter, use a calculator, use a PC, or speak a foreign language.
>
> At least nine in ten Americans agree that a country's geographic location has at least some influence on such diverse aspects of its life as the health of its population, its economic conditions, political system, and even the temperament of its people. (National Geographic Society, 1988, p. 3)

However, the report also tells us that many Americans are "seriously lacking in basic geographic knowledge and skills [and that] this is particularly true of the youngest age group tested, those between the ages of 18 and 24'' (National Geographic Society, 1988, p. 3). We are told in this connection that:

56% of American adults do not know the population of the United States.

Despite heavy U.S. involvement in Central America only half of all U.S. adults know that the Sandinistas and Contras [were] fighting in Nicaraugua.

One in three (32%) Americans cannot name any of the members of [the North Atlantic Treaty Organization,] NATO. Sixteen percent (16%) bestow NATO membership on the [former] Soviet Union, and even one in ten (11%) college graduates make this error.

Half (50%) cannot name any Warsaw Pact nations. One in nine (11%) named the United States as belonging to this alliance. [The inability of students to answer these questions correctly is at least one indication of the ephemeral nature of such ''knowledge.'']

Despite the acknowledged importance of map-reading skills, about three in ten adults cannot use one to tell direction or calculate the approximate distance between two points.

The average American adult could identify from outline maps:

- only about 4 of 12 European countries.
- less than 3 of 8 South American countries.
- less than 6 of 10 U.S. states.
- less than 9 of 16 key places on a map of the world.

Only 57% could identify England on a map of Europe. Brazil (61%) is the only South American country correctly identified by even half of the respondents. New York State was correctly identified from a map of the 48 contiguous states by only 55% (37 different states were identified as New York, from Maine to Florida, and from coast to coast).

One in seven (14%) could not identify the U.S. from a world map. This projects to 24 million adults. One in four could not identify the [former] Soviet Union or Pacific Ocean (approximately 44 million people). (National Geographic Society, 1988, pp. 3–4)

It is worth noting that the authors of this survey believe geographic illiteracy to be worsening. To prove their point, they make the claim—although it remains largely unsubstantiated—that 40 years ago, high school students, for example, could identify six of twelve European nations (compared with four of twelve reported for all adults). However, in the 18-to-24 age group sample, that figure dropped to less than three and a half, again suggesting deterioration of school instruction. However, we might also remind ourselves that high school students 40 years prior to the survey were also members of a generation that had been only too aware of the vast conflagration, with all its geographical implications, known as World War II.

The Gallup survey also compared American adults with those in Japan, Sweden, the United Kingdom, France, West Germany, Italy, Canada, and Mexico. American adults fared poorly, being matched in their ignorance only by adults from Italy and Mexico. Finally, for our purposes here, the researchers point out that the United States was the only country in the international comparison in which young adults (18–24 years) did not surpass the oldest age group tested, those 55 years and older.

In comparing these results with those of Mann's Boston School Committee and in assessing the current situation generally, several considerations need to be taken into mind. First is the obvious lack of comparison between the responding populations. In the 1845 sample, the students were on average 14 years 2 months in age. They were also a highly selected group of students, since most youngsters dropped out of school long before reaching the eighth grade. However, it is also true that when the Mann survey was replicated in 1919, when "scholars" now stayed in school longer, it was found there had been both deterioration and improvement. Compared to the 1845 sample, students scored higher on questions that emphasized meanings and relationships while scoring less well on items of fact (Caldwell & Courtis, 1925, p. 84), a difference the researchers attributed to the effect of changes in teaching resulting from the incorporation of ideas drawn from the progressive movement in education.

A second factor to be considered in making a comparison is the kind of information being sought. Do the tests, taken individually or together, reflect an assessment of knowledge—of what we might rightfully declare to be geography? Do they solicit geographic information? To answer these questions fully enough to use the results in curriculum planning, an understanding of the nature of geography and geographic thinking is obviously necessary (a matter treated in detail in the following chapter. A superficial comparison of the two tests does reveal some interesting contrasts, however. Although we must infer from the information given here the kinds of questions that lie behind the results of the survey sponsored by the National Geographic Society (1988) (the full questionnaire appears in the "Summary," p. 94) the impression is that the more recent test emphasized, to a greater degree than its 1845 counterpart, quite specific knowledge of location, and notably that kind of information that is political in nature. In contrast, the earlier test tended to place more emphasis on the relationships and configurations of physical phenomena, especially within the United States. The question begging to be answered, of course, is whether either of these emphases represents the kind of understandings that ought to be the focus of a curriculum in geographic education.

Although similar surveys over the years have also detailed what appears to be a dismal lack of geographic knowledge, no matter how it is defined, the National Geographic Society is the first to report differential scores between age cohorts and to assert that the situation is deteriorating even further than the naysayers have claimed over the years. If such a deterioration is real, then there is reason for even greater concern over the state of geographic literacy, which reinforces the urgency to understand and attempt to correct a direction none find acceptable.

## THE PLACE OF GEOGRAPHY IN THE AMERICAN PSYCHE

Unfortunately for those who would like to find a quick solution to the problem of geographic literacy, the teaching/learning process is infinitely complex. Thus, when a major issue, such as the state of geographic education, comes to one's attention, while it should be self-evident that there is no single culprit, there remains the tendency among those who would like a quick solution to adjure us that the real problem is simply that geography is not being taught. Thus, it would appear that several facts about the nature of modern geography and about its role in American life deserve consideration before we leap to the conclusion that it is all simply a matter of finding a spot in the curriculum and then discovering the will to teach "geography" as it was generally, but erroneously, taught in the past.

The first among these facts has to do with the state of geography as an academic discipline in the United States. It comes as a surprise to many that the advent of academic or professional geography—that is, the teaching of its concepts and methods in colleges and universities—is a relatively recent phenomenon. The first geography department was established in Germany in 1874 (James & Martin, 1981, p. 133), and in the United States, it was not until 1903 (Holt-Jensen, 1988). Most of the early departments were associated with geology, and between 1902 and 1914 geography assumed the role of a distinct discipline in only fifteen universities—a number that would, however, expand markedly during the period between the world wars.

As we shall see in Chapter 2, the *practice* of geography goes back to the most ancient of times, and we do indeed think of the ancient Greeks, Strabo and Ptolemy. Much later, in the nineteenth century, Alexander von Humboldt (1769–1859) and Carl Ritter (1779–1859) were also notable geographers. Even the great philosopher Immanuel Kant (1724–1804) taught a course that we would now call geography during his years as a university professor in Germany. However, these great thinkers, and all who came in the years between, engaged in geographic inquiry quite without any formal training in the subject of geography. In the United States, for example, the great explorers of the American West, people like Meriwether Lewis, George Rogers Clark, and John C. Fremont, drew, in their own unique ways, on their knowledge of surveying and other related skills to map and describe lands heretofore seen only by the American Indian. The systematic study of the earth's surface thus awaited the advent of the world of scholarship that took place during the nineteenth century, first in Europe and then in the United States.

The importance of geography's appearance as a discipline relates significantly to the fact that teachers who would teach geographic ideas require some organized knowledge of them if they are to be successful in their teaching. Obviously, this requires a large cadre of professional geographers whose interests and capabilities make possible the education of teachers in the concepts and methodologies of the discipline sufficient that they may be translated into classroom practices.

Unfortunately, while geography enjoyed a fairly rapid expansion into university, if not so much college, curriculums during the first two thirds of this century, the ensuing third has been characterized by a relative dwindling of the number of departments and a consequent reduction in the ratio of the number of trained geographers in relation to the apparent need—a ratio that was never very high. With relatively few geographers in American colleges and universities in the first instance, the decline in the number of geographers being trained over the past several decades has made it extremely difficult for teacher education programs to have access to the concepts and methods of modern geography. That in turn may well account for the evident deterioration in public knowledge of what geographic thinking contributes to our understanding of the world around us as we contrast the level of questioning in the Boston School Survey with the Gallup survey done for the National Geographic Society (National Geographic Society, 1988).

An additional difficulty has been the inability of geographers to reach agreement regarding the nature and dimensions of their field. Geography is often seen as overlapping geology particularly, but also as overlapping history, cartography, biology, and other fields in both the natural and social sciences. Due to the nature of geography this is not surprising; like history, it does not organize its concerns by virtue of a specific subject matter but rather as a consequence of its methodology. In the instance of history, the organization is chronological, in geography, it is spatial. However, it remains true that, given a relatively weak base in terms of the number of practitioners and the disagreement over its definition of the field, coupled with a relative decline in the number of institutions preparing professional geographers, American geography has been put at a particular disadvantage over the past several decades.

Thus it is that in colleges and universities that prepare teachers, there have been fewer and fewer opportunities relative to the number in most need of such knowledge—namely, candidates for the teaching profession—to have any contact with geography as a substantive field of inquiry (Boehm, Bierley, & Sharma, 1994). However, yet another complication arises.

## THE ROLE OF GEOGRAPHY IN THE SCHOOL CURRICULUM

Often there is a tendency to look to the past in the belief that it is possible to find more efficient practices which might be transferred, more or less intact, to solve today's problems. Prior to the birth of the Progressive movement in the post–Civil War period, a form of what we would today call geography was taught in U.S. schools, often with a vengeance. Those who look back fondly on the teaching of that time should be reminded that it was based on the notion of rote learning. Textbooks of the period were most often organized around a question-answer format with discussion questions scattered about, as in the example shown in Figure 1.1, which is taken from a text originally published in 1828 (Olney, 1842). Here we see the type of textbook used to assure that Boston's schoolchildren would provide the correct answers to the 1845 test used by Mann and the Boston School Committee. Since many of these students would have learned their geography from this book or others just like it, it becomes obvious that the teaching practices being advocated worked poorly, if at all.

As the century came to an end, American public education began an extraordinary period of expansion. From the newly created kindergarten (or "child's garden") to the college and university, new institutions came into being. For example, from 1890 to 1940, high school enrollments doubled every ten years. New public and private colleges and universities were founded, adding to the venerable but short list of postsecondary institutions (led by Harvard, Yale, William and Mary, and a few others established many years before). Moreover, it was during this period that the notion of providing universal education through the secondary school years and of building widespread opportunity for a postsecondary education became institutionalized in the American mind as well as in law.

The exponential expansion of knowledge since the work of Charles Darwin burst upon the scene in the mid-1800s made it now impossible for anyone to master virtually all of a field of knowledge as von Humboldt and Ritter had in geography just a few years before. Even the long-held idea that schools could teach everything essential for a citizen to know, on which instruction and the textbooks designed to achieve this end had been based, came to an end, forcing educationists to select from a growing panoply of knowledge those things thought to be of most value.

However, it was not just a simple matter of determining what knowledge might be of most worth. As the ideal of universal education, at least through the high school years, came to be seen as a viable national objective, conflict arose between those who would serve the needs of students who were college bound and those who would plan a curriculum designed more directly to serve the needs of students who were not. Complicating matters further was the growing debate over how teaching should be carried out. Traditional methods emphasizing learning as an essentially receptive activity were increasingly being challenged by the Progressives, who believed that *process* was at least as important as substance (or subject matter).

Beginning in the mid-1890s a series of committees was brought into being by the most prominent scholarly and professional societies of the day for the purpose of influencing the social science/history curriculum in American schools. Figure 1.2 summarizes the recommendations produced over a period of almost 30 years. In the earlier years of this effort, the committees were called into existence by the American Historial Association (AHA). Although a few school administrators and even one or two teachers were included in their membership, the commissions were firmly under the control of academic historians,

**Figure 1.1**
**Typical Geography Text of 1828**

8 GEOGRAPHICAL DEFINITIONS.

ts? Can you mention any thing that resembles the earth in shape? The circumference* of the earth is about twenty-five thousand miles; now if a man should travel one thousand miles in a month, how long would it take him to travel round it? The diameter† of the earth is about eight thousand miles; if it were possible to pass through the centre of it, how long would a man be at the above rate in travelling that distance? Did you ever hear of a person who had sailed round the earth or world? How much of the earth's surface is covered with water? A. About two thirds.

NATURAL DIVISIONS OF LAND.

Q. How is the Land divided?

A. Into Continents, Islands, Peninsulas, Isthmuses,

* Circumference, distance round the outside of a thing.

† Diameter, distance through a thing.

GEOGRAPHICAL DEFINITIONS. 9

Capes, Promontories, Mountains, Shores or Coasts, &c.

Q. What is a Continent?

A. It is a vast extent of land not separated by water; as America.

Can you show me America on the map of the World? What is the northern part of it called? What is the southern part called?

Q. What is an Island?

A. It is a portion of land surrounded by water; as Australia,* Borneo, Iceland, Cuba, &c.

Did you ever see an Island?

Q. What is a Peninsula?

A. It is a portion of land almost surrounded by water; as Africa, South America, &c.

Can you show me Africa? Is it entirely surrounded by water? What is the narrow neck of land called which connects it with Asia?

A. Isthmus of Suez.

Q. What is an isthmus?

A. It is a neck of land which joins a peninsula to a continent, or unites two parts of a continent; as the Isthmus of Suez, Isthmus of Darien, &c.

Can you tell me where the isthmus of Darien is? What does it unite? What body of water is north of the isthmus of Darien? What body of water is south of it?

Q. What is a Cape?

A. It is a point of land extending into the sea; as Cape Horn, Cape Farewell, Cape of Good Hope, &c.

Can you show me a Cape on the map?

Q. What is a Promontory?

A. It is a high point of land extending into the sea; as the southern part of South America, Hindoostan, &c.

If a mountain extended into the sea, what would you call the end of it? Why?

Q. What is a Mountain?

A. It is a vast elevation of land; as the Andes, Alps, White Mountains.

* Australia has, until recently, been called New Holland.

Illustrated here are the flyleaf and two pages selected at random from the main body of the Olney text. Note the rigid question-answer format, typical of the books of that time (early 1800s), in the illustration above. The flyleaf (facing page) conveys the then-common emphasis on the European connection with the United States. The presumed superiority of that relationship is given added credence by making invidious comparisons through commonly held stereotypes of societies outside that august orbit.

**Figure 1.1 (continued)**

**Figure 1.2**
**Summary of Recommendations for Curriculum Development**

| Name of Committee | Date | Membership of Committee | Third Grade | Fourth Grade | Fifth Grade | Sixth Grade |
|---|---|---|---|---|---|---|
| Committee of Ten N. E. A. A. H. A. | 1892 1894 | 7 Professors<br>3 Principals | | | Biography & Mythology | Biography & Mythology |
| Committee of Seven A. H. A. | 1896 1899 | 6 Professors<br>1 Principal | Stories from the Illiad, etc. | Biography | Greek and Roman History | Medieval & Modern History |
| Committee of Eight A. H. A. | 1905 1909 | 4 Professors<br>2 Superintendents<br>2 Teachers | Heroes of other times. Pictures of scenes and persons of various ages | American history; exploration to Revolution. Historical scenes and persons in early American History | American History; Revolution to the Civil War. Historical scenes & persons in later American History | European background |
| Committee of Five | 1907 1912 | 4 Professors<br>1 Principal | Chiefly Biography: Civics to be taught Grades 1 to 8 | | | |
| Social Studies Committee N. E. A. | 1914 1916 | 5 Professors<br>2 Superintendents<br>10 Teachers<br>4 unclassified | | | | |
| Committee of Seven A. P. A. | 1911 1916 | 6 Professors<br>1 Superintendent | Civic virtues | A study of simple community activities<br>Little textbook work | | |
| | | | | The Making of the United States | | |
| Committee on History & Education for Citizenship, A. H. A. N. E. A. | 1918 1921 | 6 Professors of History<br>1 Professor of Education<br>1 Superintendent<br>1 Teacher | Discovery and Exploration | How Englishmen became Americans 1607-1783 | The United States 1783-1877 | The United States. 1877 to date, 1/2 yr Civics 1/2 yr |

A. H. A. , American Historical Association
N. E. A. , National Education Association
A. P. A. , American Political Science Association

| Committee | Seventh Grade | Eighth Grade | Ninth Grade | Tenth Grade | Eleventh Grade | Twelfth Grade |
|---|---|---|---|---|---|---|
| Committee of Ten N. E. A. A. H. A. | American History and Civil Government | Greek and Roman History with Oriental Connections | French History with background of Medieval and Modern History | English History with background of Medieval and Modern History | American History | One Special period & Civil Government |
| Committee of Seven A. H. A. | English History | American History | Ancient History to 800 | Medieval & Modern History | English History | American History and Civil Government |
| Committee of Eight A. H. A. | Early American History 1500–1789. Still more Civics | Later American History 1789–1909 (Also some Modern European History ) | | | | |
| Committee of Five | | | Ancient History to 800 (Econ., Pol. & Soc.) | English History with Continental Connections to 1760 | Modern Europe English Connections since 1760 | American History and Government (Separately or ratio 3:2) |
| Social Studies Committee N. E. A. | Geography European History, and Community Civics | American History Community Civics and Geography Incidentally | Political, Economic, & Vocational Civics with History incidentally | Ancient and Medieval History to 1700 (1 yr.); Modern European History (1/2 or 1 yr.); American History since 17th Century (1/2 or 1 yr ); Problems of American Democracy (1/2 or 1 yr.) | | |
| Committee of Seven A. P. A. | Community Civics (Emphasis upon functions, but some treatment of machinery of Government.) | | | An advanced course in Civics (Report does not state in which year to be offered, nor whether 1 or 1/2 yr.) | | |
| Committee on History & Education for Citizenship, A. H. A. N. E. A. | American History in its World Setting | | | The Modern World | | |
| | The World before 1607 (Including Spain in America | The World since 1607 with emphasis on Economic & Social History of the United States | Community & National Activities, incl. Commercial Geography, Civics, Socio., & Economic History | Modern European History since 1650 | American History during National Period | Social Economic & Political Problems & Principles |

*Source*: Rugg (1926), pp. 48–49. Reproduced with the permission of the National Society for the Study of Education.

whose self-interest showed clearly in the forthcoming recommendations. The professional association for geographers in the United States comparable to the American Historical Association, the Association of American Geographers, was not established until 1904, far too late to influence these deliberations, which were clearly designed to influence the content of the school curriculum on a national scale. As well, of course, while the need to consider the needs of non–college-bound students was ostensibly a primary reason for calling the committees into being in the first place, their makeup strongly suggests where the priorities lay.

As the search for a perfect curriculum continued, other groups, most notably the American Political Science Association and National Education Association, became involved in the process of recommending a subject matter content aimed at preparing future citizens for participation in civic life. Although even these groups concurred in the importance of history, and particularly those aspects that might serve as the raison d'être for an emerging democratic society, they were at least more attuned to the idea of studying topics in something other than a strictly chronological mode. This was especially true for the committee established by the National Education Association (NEA) in 1912 (Saxe, 1991, p. 144). Influenced by the progressivism of the period, and especially by the educational philosophies of John Dewey, William James, and other contemporaries who perceived the process and purpose of educating as being at least on a par with the acquisition of knowledge, the work of this committee was carried out under a new rubric, that of the *social studies*. Here we see, not only a major departure from the recommendations of prior committees, but also an attempt to redefine an area of the curriculum previously dominated by the notion of separate disciplines, most notably, of course, history (generally now thought of as part of a broader scheme generally ascribed to as the humanities rather than as a social science).

For a comprehensive history of the social studies I refer the reader to David Warren Saxe's *Social Studies in Schools: A History of the Early Years* (1991) for the most comprehensive treatment of a curriculum area that continues to confuse and perplex professional educationists and the public alike. Suffice it to say here that in those earlier years, *social studies* was not perceived as a definition of a subject matter but as an approach to the utilization of those subject matters of the social sciences, and including history and geography, for a particular purpose. That purpose was to draw on knowledge generated by these so-called disciplines in such a way as to provide students, not simply with socially useful skills and information, but also with the will to participate in society with the intent of its betterment. ''Social studies'' was, quite literally, the plural form of ''social study,'' meaning that its aim was to engage students in studies that promoted the social good.

Reviewing the work of the various curriculum committees formed during this period reveals the power of the ''history lobby,'' even in the face of what today would be judged a more enlightened view of how the educational process might be carried out most effectively, for then as now it enjoyed an ubiquitous presence throughout both elementary and secondary school curriculums. Moreover, while the influence of the American Historical Association in setting this stage should not be underestimated, it is also true that history, meaning American history and its antecedents, assumed an importance in its own right as a kind of social glue cementing society's past with its present as well as its future. Geography had no advocate the equivalent of the AHA, nor did it exude the power inherent in the subject, which could make a fairly direct claim of having uniquely patriotic merit. Geographers were few and far between as well, and those who argued for curriculum reform were without guile when they suggested that the study of geography be limited to the accumulation of knowledge about the physical world.

## Curriculum Reform and the Scientific Movement in Education

The emergence of faith in the powers of the natural sciences in the mid- to late 1800s brought with it the belief that, not only would the natural world yield to human attempts to control it, it would help to explain human behavior as well. And while this idea, called scientism, did not come into full flower as far as the processes of schooling and educational processes generally until the 1920s, it then indeed began to be applied with a vengeance. The so-called scientific movement in education was a time for applying what was thought to be a scientific methodology to educational questions of all kinds. It is perhaps epitomized in the idea of the intelligence quotient, which is essentially a figure arrived at when a test score purporting to measure a person's mental age is multiplied by 100 and then divided by the individual's chronological age. Invented in the form we know it in the United States by Lewis Terman (1877–1956), a professor of psychology at Stanford University, it was not even an accurate shadow of the test on which it was originally based. The originator of the idea that intelligence might be estimated by devising a test of intellectual capacity was Alfred Binet (1857–1911), director of the psychology laboratory at the distinguished Sorbonne, University in Paris. But Binet's intent was quite the opposite of Terman's. His interest was in identifying mentally retarded pupils in the French schools so that they might be singled out for special attention. He sought to do this by devising a series of graded tests based on general skills and abilities not generally taught in school. Terman, moving in quite the opposite direction, sought to identify the entire range of "intelligence," in the process assuming that it was a generalized capacity affecting every aspect of life, including the academic. Intelligence was also considered to be fixed, and thus, when determined, quite literally might be used to assign children to their proper station in life. Expectations thus became the reality, and students were to be treated accordingly.

Where Binet had devised a test to be administered to the individual child, Terman wanted to test everybody; his goal, accordingly, was to devise a means by which large numbers of subjects could be tested at one time, or at least in a more efficient manner than the one-on-one protocal followed by Binet. Yet in all fairness, it must be said that Terman was not acting alone. The test with which he would later be closely identified resulted initially from a group effort (including most notably H. H. Goddard and Robert M. Yerkes) designed during World War I to screen recruits. Having had literally tens of thousands of soldiers on whom to practice, Terman set out to develop a test for general use.

With refinements suggested from this experience, Terman produced the first versions of what would be called the Stanford-Binet Test of Mental Ability. This has been followed, not only by numerous revisions, but as well by numerous other tests purporting to produce a similar measure (the scores for which have consistently been standardized on the Stanford-Binet). I mention this in some detail because the Stanford-Binet and its underlying assumptions (that intellectual content is quantifiably measurable and an accurate marker of what it presumes to evaluate) provided the basis for developing tests designed to measure the content of human minds (see Gould, 1981, for a comprehensive discussion of the development of the intelligence test). Scores from these tests have also, periodically, been used to suggest racial superiority/inferiority, particularly to suggest the intellectual superiority of white students over students of color (see, for example, Jensen, 1980; Murray & Herrnstein, 1994).

Following the notion that intelligence might be measured with a high degree of accuracy, the 1920s would give birth to tests in a number of the subject matter areas of the curriculum, from reading and arithmetic to the subject matter side of the social studies, including geography. Like tests of intelligence, these would be continuously revised. The most authoritative reviews of standardized tests, the *Mental Measurements Yearbooks*, were published over a period of more than 40 years by O. K. Buros and associates, (Buros, 1975). Perusal of these yearbooks will reveal to the interested reader the origins and long tenure of a very large proportion of the tests still in use in American schools (no other country relies on standardized testing to this extent). The reviews contained in the yearbooks provide the reader with repeated cautions about these tests. Nonetheless, despite these and other warnings of the inadequacy of both intelligence and achievement tests, if anything, faith in their results is now stronger than at any time in the past, a consequence, it would seem, of a growing general public distrust in the nation's schools.

An irony in this development is that, even as the overall scores are assumed to be diagnostic, subscores are also commonly thought to be accurate measures of weaknesses within the overall frame being assessed. To further compound the problem, subsets of questions are often evaluated as indicators of specific abilities assumed to be part of the overall assessment question. Problems of reliability within the sub-sets (the internal reliability) are consequently put to one side by test makers, who are happy enough to obtain high overall test reliability (i.e., subsequent administrations of the test produce similar scores), even though this raises further questions about the more important factor of validity, or whether the test actually assesses what it purports to evaluate.

Standardized tests based on subject matter from the social sciences share this problem of questionable validity. In the findings of the National Assessment of Educational Progress (U.S. Office of Education, 1990), the extent of students' knowledge of geography is based on answers to only a subset of questions, and, moreover, on a subset of the entire population taking the test. In standardized testing generally, this is not only an accepted procedure, it is the only practical solution to the problem of keeping students' noses to the evaluation grindstone. But it does raise the question of whether the picture provided by the test score is an accurate one. Consequently, how much time schools should spend on such evaluation practices has, over the years, become an issue. Some educationists would argue that testing halts the learning process since the activity itself emphasizes recall of what has already (presumably or hopefully) been learned.

Although standardized testing was a major development as educationists sought to bring the presumed benefits of the scientific method to bear on school problems, there was another series of efforts, which experienced quite a different outcome. I refer here to the continuation of earlier attempts to redefine the social studies/social science curriculum. Beginning in the 1920s, there developed a growing dissatisfaction with the traditional academically oriented curriculum, which was illustrated repeatedly in the attempts of the various curriculum committees described earlier in the chapter. As high school enrollments burgeoned, the tinkering that seemed to characterize those efforts was put aside for a more rational (or scientific) method of selecting curriculum content. I emphasize the matter of intent, for it should be understood that, as today, theory and practice differed. However, it can fairly be said there did occur serious attempts to wed these two seemingly disparate concepts (see, for example, Part 2 of the 22nd yearbook of the National Society for the Study of Education Yearbook, *The Social Studies in the Elementary and Secondary School*, 1922).

If schools were to educate a ''new'' generation—one committed to social progress and civic participation, as the Progressive voices of the time appeared to demand—the sentiment was to look more toward empirical than philosophical or historical sources for curriculum making. In the social studies, this meant looking toward more contemporary problems and issues as sources for curriculum content. It seemed logical, then, if the curriculum were to reflect current social needs and concerns, to turn to more contemporary sources. One approach to the problem was to analyze the contents of contemporary documents such as newspapers. Another source for content selection was found in analyses of topics that the ''frontier thinkers'' had included in their writings for answers to the question of what to teach (see, for example, Hockett, 1927; Billings, 1929; Rugg, 1926).

However, social studies educators would soon learn what a dangerous ''subject'' their field really was. Drawing on the notion of developing a more contemporary approach to the social study, Harold Rugg, a professor at Teachers College, Columbia University, developed a series of textbooks for junior and senior high schools under the title, ''Problems of American Culture'' (see Rugg, 1931). Published in the early 1930s, they became instantly popular and remained so until critics got wind of the contemporary nature of the subject matter, whereupon they were systematically withdrawn from use, and in a few instances, even burned in protest. The deepening economic depression spread its uncertainties into other aspects of American life, and reality was not what a distraught public considered proper subject matter for students. Perhaps misinterpreting the public mood, there followed a series of illustrated newspaper-like monographs for individual student use, which were later bound together and distributed by the Americana Corporation, publisher of the prestigous *Encyclopedia Americana*. Called ''Building America'' (1938–1946), the series was put together by writers under the direction of some of the most influential social studies educators of the time under the sponsorship of the Department of Supervision and Curriculum Development of the NEA. ''A series of pictorial study units on modern problems,'' which treated current social problems such as strikebreaking, breadlines, support for the aged, unemployment insurance and other socially sensitive topics, ultimately bringing about the expected strong negative public reaction (Foshay, 1990, pp. 29–30). In retrospect it is not surprising that these materials served to inflame the passions of those who believed, as many do today, that children should be seen and not heard and that the educational process itself is best conducted, if not quite in total silence, then certainly without controversy.

It was nevertheless during this period—the late 1920s and the 1930s—that the social studies curriculum pattern now common to American public schools had its genesis. Its major architect was Paul R. Hanna (1902–1989), and its structure would find its genesis in the first statewide effort at school curriculum improvement, which was undertaken in the state of Virginia in the late 1920s, while Hanna was a doctoral student at Teachers College, Columbia University. (He would later become a professor of education at Stanford University, chairman of the editorial committee of ''Building America,'' social science editor for *World Book Encyclopedia*, and a senior editor for Scott-Foresman Publishers, among other positions of influence where curriculum development in the social studies was concerned.) As a consultant to the social studies committee of what would be called the Virginia Study (Virginia Board of Education, 1934), he formulated a design for the social studies curriculum that would serve as the centerpiece of his professional interests.

Influenced by the progressives of the day—Teachers College was at the center of the ''new'' thinking in the social studies—Hanna eschewed the traditional subject curriculum. Instead of basing the social studies on the customary academic disciplines, he further developed what are technically termed *scope* and

**Figure 1.3**
**Basic Human Activities Superimposed on Expanding Communities of Men**

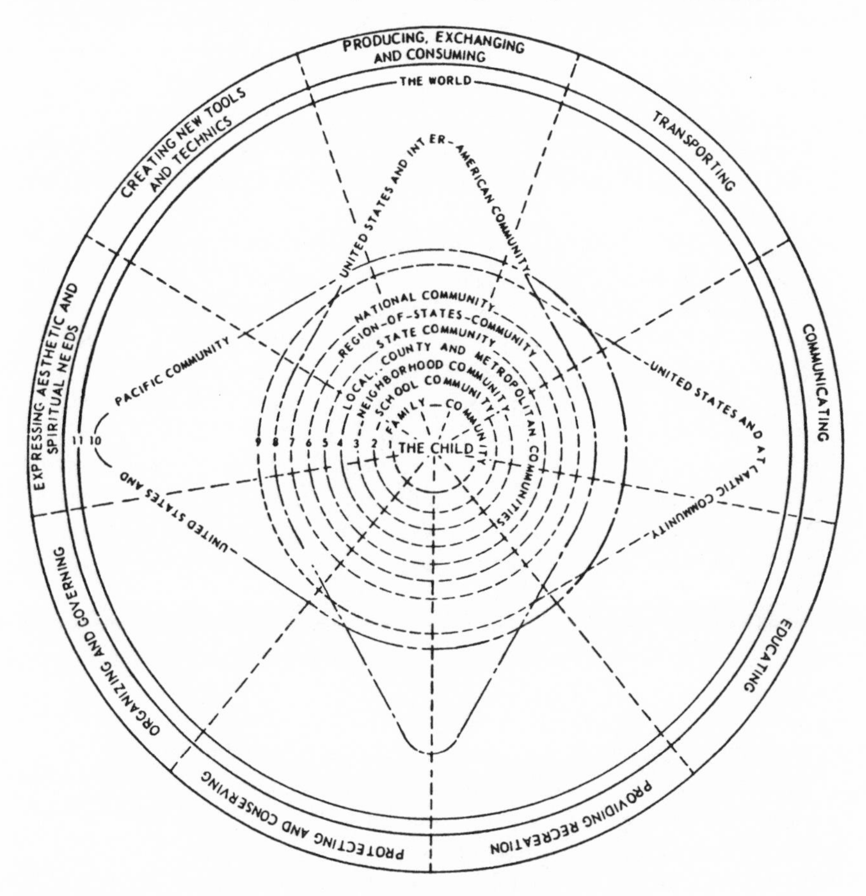


*Source*: Hanna et al. (1966). Copyright © 1966 by Houghton Mifflin Company. Used with permission.

*sequence*, involving a design that delineates both a series of topics over time and the breadth of those topics at any point along the graded continuum.

The Hanna-designed concepts of scope and sequence are illustrated in Figure 1.3 (Hanna et al., 1966). The figure depicts his conception of the expanding communities, the study of which provides the sequence in which students would, theoretically at least, gradually grow in their knowledge of the world by beginning with the local community and then gradually moving out to the larger world. These are depicted in the center of the chart. The scope, or breadth of such studies is defined by nine ''human activities,'' which were presumed to describe the breadth of behaviors engaged in by all humans, whether in a primitive or a more advanced state. Hanna shows these items around the circumference.

Hanna, and the many other educators who accepted the basic premise of this design of moving from one's immediate environment to the larger environment of the state, nation, and ultimately the world, perceived the customary social science disciplines as contributing to the topics being studied, but not as subjects that should be considered in their own right. Whether accepted only tacitly or not at all by current school practitioners, the design has long been the dominant pattern of the social studies curriculum in America. However, it is interesting to note that even Hanna had to give way on what turned out to be a major, untouchable point: The tradition of teaching history, particularly American history, could not be avoided. Thus, a major deviation from the logic of the so-called expanding communities concept appears, quite irrationally, at the fifth, eighth, and eleventh grade levels. Aside from this deviation, Figure 1.4 suggests the role the traditional disciplines were seen to play in furthering the expanding communities concept. Advocates argue the concept's psychological soundness, saying that it takes the child from the local community, from the part of the world said to be most familiar since it is the first one to be experienced directly, on a course gradually outward and, increasingly, physically more distant, and therefore more difficult to comprehend. The basic human functions, it has been claimed, provide the necessary assurance that students will become familiar with the many and varied activities in which all people engage (to a greater or lesser degree, depending on their particular talents and resources).

While the psychological soundness of the expanding communities concept may have been a more acceptable argument in times past, its continued use as a base for making curriculum decisions depends most directly on the seeming, rather than actual, logic of its scope and sequence. This pseudo-logic has, in fact, now been disturbed by numerous discoveries regarding the development of social, spatial, and chronological concepts and by information about reading and language development generally, which I believe requires its reexamination as a sound organizing instructional principle.

The influence of Hanna's rationale for a social studies curriculum cannot be underestimated (see, e.g., Hanna, 1987). Although more than 60 years have passed since its initial formulation, and while the notion of *basic human activities*, although it influenced textbook writers, never did take hold in the classroom, the topical sequence he suggested has very largely assumed a life of its own. Although the Hanna plan has come to dominate curricular organization in the social studies, at the same time it has served as a lightning rod in an ongoing argument that erupted following World War II, when the notion of progressivism in education came to an abrupt end. The Korean War, accompanied and followed by a number of international crises and social problems, brought forth a number of critics who believed that process had outstripped something much more important, that is, the idea of substance, or knowledge of subject matter. With books such as newspaperman Rudolf Flesch's *Why Johnny Can't Read* (1955) and historian Arthur Bestor's *The Restoration of Learning* (1955), a steady drumbeat of criticism began, to be interrupted briefly by the launching of *Sputnik*

**Figure 1.4**
**Role of the Traditional Disciplines in Furthering the Expanding Communities Concept**

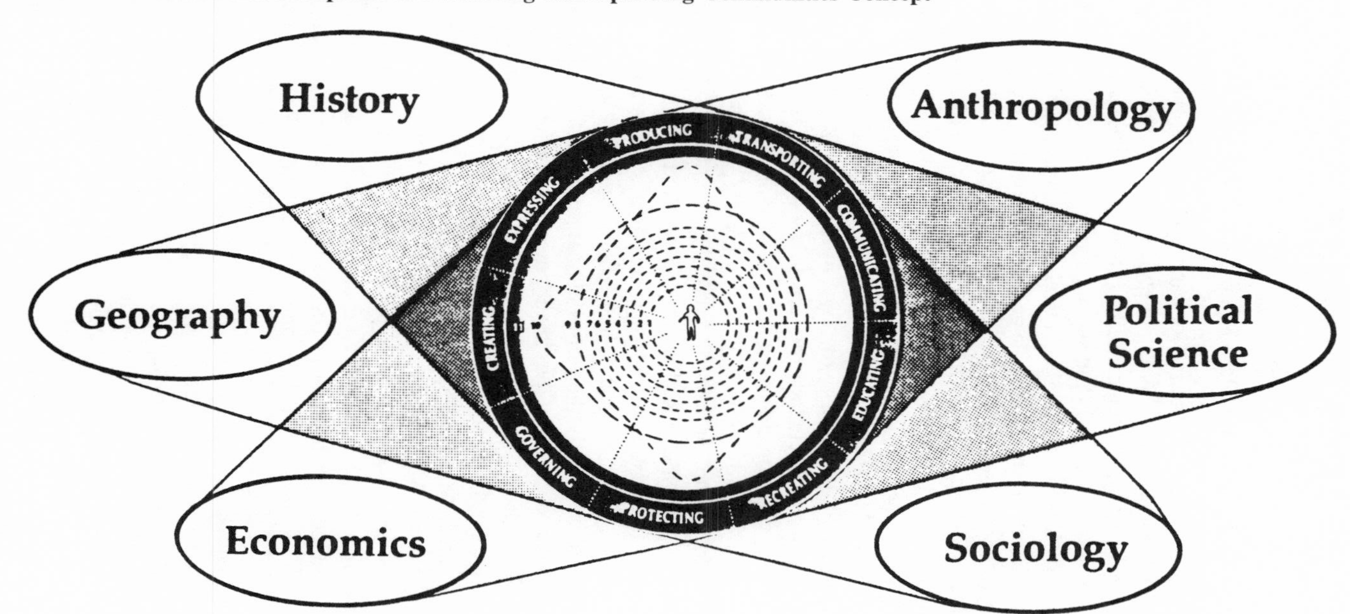


*Source*: Hanna (1987). Reproduced with the permission of the publisher, Hoover Institution Press. 

in 1958, when a huge gap in learning in areas where substance had always been a priority—in mathematics and science—was presumed to have appeared on the horizon.

## Curriculum Reform in the Post-Progressive Era

The story of geography in the school curriculum would therefore not be complete if we were to overlook this relatively recent (occurring during the 1960s and on into the 1970s), although failed, effort to achieve a better balance between substance and process in the social science curriculum (Saxe, 1991). The shock of *Sputnik* led to an intense, but relatively brief, period directed toward improving the content curriculum, especially in the sciences and mathematics but also where social studies was concerned. Jerome Bruner, a psychologist at Harvard University, emerged during this period as the spokesman for what was to be called *discovery learning* (Bruner, 1961). It was a point of view that emphasized both content and process, although Bruner placed a renewed emphasis on process. He asserted that "any subject can be taught effectively in some intellectually honest form to any child at any stage of development" (p. 33). From this grew demands that teachers obtain a better grasp of the intellectual material undergirding their instruction. Simultaneously, Bruner argued that the process of discovery was more compatible with how learning actually proceeds than traditional notions allowed, that is, notions in which the mind was perceived as an empty vessel waiting to be filled with information. Instead, students were to be encouraged to approach their studies much as the mature scientist might. Large sums of state and federal funds, as well as support from the private sector, especially the Ford Foundation's Fund for the Advancement of Education, were devoted to the development of curriculums at both the elementary and secondary levels in pursuit of this objective. Notable among them in the sciences were the Biological Sciences Curriculum Study (BSCS) and the Chemical Education Materials Study (CHEMS) (Grobman, 1970) for secondary students; the School Mathematics Study Group (SMSG) (Tanner & Tanner, 1975) and the Science Curriculum Improvement Study (SCIS) (Karplus, 1967), which focused on the early years, and, in the social studies, "Man: A Course of Study" (MACOS), a particular project of Bruner and his associates, for upper elementary school (grades 5 and 6); and, especially important for our purposes, the High-School Geography Project (Goodlad, 1964).

The High-School Geography Project differed from those others that sought to redefine the content of the high school curriculum because of certain problems endemic in geography itself. Because of its history, or lack thereof, as a recognized emphasis within the social studies curriculum, there was the additional need to clarify once and for all a substantial place for geography within the curriculum. To that end, professional geographers sought to develop a more convincing rationale and content emphasis. As Goodlad points out in his review of the project (1964, p. 44), the lack of a clearly defined and established position in the high school required a different approach to the problem of curriculum reform in geographic education. Instead of developing instructional materials, several experimental teaching projects devoted to the development of geography courses were initiated. In addition, a group of cooperating teachers, not supported by outside financial aid, attempted to develop course outlines and help in the refinement of various project activities.

For various reasons, this and the many other attempts at curriculum reform that were launched during the late 1950s and on into the 1960s, and early 1970s failed to gain a foothold. For example, in the case of BSCS, the major cause was the apparent failure of students to perform significantly better than those enrolled in conventional programs. (A continuing problem for any "experimental" program is that such comparisons depend on tests devised to evaluate the acquisition of subject matter taught in a traditional curriculum. New or experimental curriculums generally seek additional goals which, while they may be met to a greater or lesser degree, are not included in traditional forms of assessment.) SMSG failed due to the inability of teachers (and parents) to understand the mathematical vocabulary and implied concepts undergirding the curriculum. Then, the broken front line of attack of the High-School Geography Project eventually fell victim to its own disorganized plan for curriculum reform, in which the need to define the place of geography in the school conflicted with the need to find ways to integrate geographic concepts into the total curriculum. In the case of MACOS, politics reared its ugly head, triggering a general collapse of what had been anticipated would be a new and powerful way of improving schools on a macro scale (Boyd, 1979).

MACOS was conceived as an integrated curriculum which incorporated a wide range of instructional materials for student use along with aids designed to increase the content background of the teacher. It deserves particular attention for it illustrates how particularly vulnerable are the social sciences for bestirring value conflicts within American society, where heterogeneity rather than homogeneity of opinion abounds. The content of the MACOS curriculum was derived from all the social sciences, but particularly anthropology, in which the compare/contrast teaching processes focused on often-controversial topics. Because of its content—a most frequently cited example of which was a unit of study of Eskimo life that critics viewed as excessively and brutally realistic and as foreign to values enshrined in the American culture—MACOS drew its critics as soon as it began to be disseminated beyond the experimental curriculum development stage. Moreover, because its genesis was largely funded by the federal government, it quickly drew the ire of the U.S. Congress. In a detailed analysis of what came to be known as the MACOS controversy, Schaffarzick notes that "protests ranged from complaints about perceived attempts by scholars and experts to shape local values, to criticisms of MACOS' message of relativism, and charges that the course was a 'pernicious attempt to spread the religion of secular humanism' " (1979, p. 9).

I mention the MACOS debacle in some detail because it illustrates how dangerous topics in any of the social sciences, when developed beyond the level of innocuous information, are often judged from particular political and social perspectives quite apart from their educational value. While many viewed it as an exceptionally high-quality curriculum development effort in the social studies (an aspect of the schools' work often derided as boring and ineffective), its value-laden content provoked hysterical outbursts, the upshot of which was the withdrawal of federal activity in direct efforts to improve the nation's schools through curriculum reform. (More recent attempts have centered on instituting standardized tests and the setting of ''national standards,'' in the process bypassing the question of how something might best be taught for the easier task of saying what must be learned.)

## The Current Circumstance

The curriculum development debacle of the 1960s and early 1970s saw the abandonment of, not just geography and history, but the social science curriculum generally. Responding to the criticism that the home and the church were the only legitimate institutions responsible for shaping values, local school units followed public pressure to emphasize the transmission of facts and the development of skills.

Geographic education, and indeed the social studies generally, were to suffer grievously as a consequence. The ''back to basics'' movement was combined with a taxpayers' revolt to bring about major changes in what would be taught. In the elementary school, this meant the curriculum became increasingly skewed toward the teaching of reading and computational skills. In high school, it meant an increased emphasis on ''tracking,'' or segregating students into groups with specialized missions: those who were college bound (headed by an elite of ''Advanced Placement'' students), the so-called ''General Education'' students, and those enrolled in Voc Ed, or ''Vocational Education.'' Later, this also involved the organizing of ''magnet schools,'' which not only emphasized specializations from the arts to the sciences but began to be considered institutions where students could get a leg up on a career.

By the 1980s it had become clear that the most recent ''back-to-basics'' movement had failed to bring about significant school improvement. Several publications commissioned by federal and foundation sources documented what was perceived as a growing national educational crisis. In an attempt to better understand American schools and to help lay the groundwork for school improvement, a number of publications called attention to the problems of the schools, including the widely circulated *A Nation at Risk: The Imperative for Educational Reform* (National Commission on Excellence in Education, 1983). One of the most detailed of the studies enumerating the status of American schools was conducted by John I. Goodlad concurrently with the meetings of the commission that had issued the rather polemical and certainly political *A Nation at Risk.* Just one year following the government-sponsored report, Goodlad published a broad-scale study in which he described and analyzed a nationwide, reprentative sample of elementary and secondary schools (Goodlad, 1984). Not surprisingly, he discovered that ''geography'' had been almost completely subsumed under the rubric of the social studies. Commenting on its emphasis within this larger rubric, he reported that:

> The curriculum at the elementary level was amorphous, particularly in the lower grades. Many first-and second-grade classes put together the themes of understanding self and others with discussion of the family and the community. There were more field trips—to community resources and facilities—than occurred later. The intent, apparently, was to begin close at hand, with oneself, and expand one's understanding of the immediate environment. (Goodlad, 1984, p. 210)

Teachers ''quite consistently'' reported that they endeavored to teach map skills at the intermediate grades (4–6) and that they incorporated the ''themes'' of history, geography, and civics at grades 5 and 6, ''mostly in the context of the growth and development [history] of the United States'' (p. 210).

There appeared to be a great deal more uniformity in the social studies curriculum at the junior high school level. U.S. history, world history, and world geography, frequently accompanied by state history, were commonly found. Goodlad reports: ''Teachers noted the expectation that their students would acquire map skills, the ability to take notes, proficiency in the use of dictionaries and encyclopedias, and the skills of oral and written expression'' (1984, p. 211). At the senior high school, ''the basics'' of social studies were found to be American history and government. Beyond these ''courses,'' electives were offered in a smorgesbord including economics, sociology, law, anthropology, psychology, world history, the history of the state, world cultures, human relations, current events, and the history and/or geography of a variety of other countries (p. 211).

Looking beyond what schools claim their curriculum to be and what teachers say they emphasize, an even more disturbing picture begins to emerge. For instance, following rather consistent reports over the years, Goodlad and his associates found that students reported a dislike for the social studies. At the upper elementary school level, in fact, they liked social studies less than any other subject (Goodlad, 1984, p. 212). Similarly, at the higher grade levels,

> [While] the topics commonly included in the social sciences appear as though they would be of great human interest[,] . . . something strange seems to have happened to them on the way to the classroom. The topics of study become removed from their intrinsically human character, reduced to the dates and places readers will recall memorizing for tests. (p. 212)

Referring to prior studies that also reported general student dislike for the social studies, Goodlad and his associates found that, while topics in this area of the curriculum have consistently enjoyed high ratings, social studies as a school "subject" was rated relatively low in interest among the several curriculum fields. Nor did they find, in their study, "much inclusion of global or international content. Over half of the students believed that foreign countries and their ideas are dangerous to American government" (p. 212).

> Many of the teachers in [Goodlad's] sample appeared not to have sorted out the curricular and instructional ingredients of a social studies program designed to assure understanding and appreciation of the United States as a nation among nations and its relationship to the social, political, and economic systems of other countries. (p. 213)

Goodlad concludes:

> [T]he preponderance of classroom activity involv[es] listening, reading textbooks, completing workbooks and worksheets, and taking quizzes—in contrast to a paucity of activities requiring problem solving, the achievement of group goals, students' planning and executing a project, and the like. . . . Indeed, in many ways, instruction and learning in the social studies look more like instruction and learning in the language arts (without the emphasis on mechanics) than in the social sciences. It appears that we cannot assume the cultivation of goals most appropriate to the social sciences even when social studies courses appear in the curriculum. (p. 213)

More recent data give us little encouragement that the rekindling of interest in geography as a school subject will bear much fruit. In fact, it seems to raise even more difficult questions regarding the assumption that what is taught in a deliberate fashion is also learned to a reasonable degree. In a 1990 report, *The Geography Learning of High-School Seniors* (U.S. Office of Education, 1990), which was prepared by the Educational Testing Service (ETS), the question of the relative effectiveness of what little teaching of geography there appears to be is once more called to attention. Conducted as a part of a larger study, the National Assessment of Educational Progress, it involved slightly over 3,000 students randomly selected from a much larger population, which had been drawn from a nationwide sampling of 1,500 schools. The students responded to questions in four areas: knowledge of locations, using the skills and tools of geography, understandings of cultural geography, and understandings of physical geography. Summarizing the findings, the authors conclude:

> Most students did not demonstrate an understanding of the basic concepts of physical and cultural geography, and many did not correctly identify the location of major countries, cities, and landmarks. Further, many of the students did not seem to understand that maps can be used to derive all kinds of information about the world, rather than simply to find places. (p. 7)

However, besides yet another study iterating a disappointingly low command of geographic understandings, what we see here is documentation of a disturbingly evident *lack* of relationship between teaching and learning. In presenting the data, student responses were separated among those who said they had studied "a lot," studied "some," or had had "little or no" study of geography (32, 54, and 14 percent, respectively). Although there were some regional differences, "on average, there was no significant difference in assessment performance between those reporting very little or no study and those who had at least some exposure to the topics" (U.S. Office of Education, 1990, p. 50). Particularly noteworthy is the lack of a clear-cut distinction between students who said they had studied geography "some" and those who claimed "a lot" of study. *Additionally, students reporting having taken geography in their senior year performed slightly worse than those taking it in one or another of the three previous years* (p. 52). An analysis of subgroups revealed that:

Males were more likely than females to report taking geography in high school. White students were no more likely than their Black or Hispanic counterparts to report having taken a course [although the latter two groups scored less well, with blacks scoring consistently below Hispanics]. High school students living in the West were more likely than those from other regions to report geography course-taking. . . . Students in vocational/technical high-school programs were as likely, if not more so, to state that they had taken a geography course as students either in academic or general programs.

Across the subpopulations studied, twelfth graders who reported they had taken a high-school geography course typically performed no better in the assessment than those who had not taken a course in the subject. (p. 53)

Because "geography" is more often than not incorporated in other "social studies" courses, comparisons were made to see what differences might be found when students had come across geographic concepts under these conditions. The results here were equally surprising and unsettling: students reporting that their geography instruction had been part of their U.S. history course:

demonstrated higher average proficiency in geography than their counterparts who had not taken such a course. Students who reported that geography was taught as part of earth science or courses other than U.S. history [world geography/history; physical geography/earth science; or other geography courses] generally had similar proficiency in the subject, whether or not they had taken those courses. (p. 55)

That this may not be a one-of-a-kind finding is suggested by a more recent study assessing American college students' knowledge of geographic concepts and understandings (Eve, Price, & Counts, 1993–1994). Drawing on a sample of 313 students in general level English and sociology courses at a major southwestern public university, the relationship between teaching and learning was called into question even further than in the 1990 ETS report on high school geography. When asked to indicate the last time they had taken a geography course—in elementary school, middle school, high school, or college—the college students' answers revealed no correlation between course work taken and geographical knowledge. ''In fact,'' the researchers report, ''the *opposite* of what was expected was found in some cases. Those who had taken their last geography class in elementary school did better than those who had taken one or more classes as recently as middle school, and almost as well as those who had taken geography classes in high school or college (p. 15). Overall, ''there was no discernible correlation between the number of geography classes taken and geographic ability [as measured on the test]'' (p. 15).

It is, consequently, not difficult to imagine an increasingly dark scenario. On the one hand we have social studies, a curriculum area currently assigned with the primary responsibility for teaching geographic concepts, where teaching has long been recognized as relatively ineffective as well as one that students find boring and unappealing. On the other hand, we observe an increasingly powerful media of communication (television, the Internet worldwide communication system) with a growing capability of distorting reality and whose management and direction, with only a few exceptions, have no direct link to, or even interest in, ''education'' as an organized and purposeful social activity.

## Academic Geography versus School Geography

Central to the issue of finding an appropriate role for geography in American life has been the search to identify the differences that separate academic geography—geography as it is taught in colleges and universities—from geography as it should be taught in our elementary and secondary schools. The distinction most frequently argued between the two approaches has centered on the notion of professional preparation at the collegiate level and, as a product of instruction in our elementary and secondary schools, geography as knowledge about the human and physical aspects of human occupancy of the earth's surface.

But what should be the ''content'' of the *geography* taught in schools? Early on, of course, the textbook was the primary determiner for this question. During the Progressive Era, with the formation of what have been called the *social studies*, geography became subsumed under this rubric and, according to some, lost its identity as attempts were made to integrate geographic concepts into a broader scheme. However, we would be fooling ourselves if we thought this movement affected even a majority of teachers. As Cuban has pointed out in *How Teachers Taught: Constancy and Change in American Classrooms, 1890–1990* (1993), traditional methods have persisted despite cries for innovation and change over the years. Teachers consequently still look beyond their own intellectual resources to determine what will be taught in the classroom.

Where, then, should teachers turn to select an appropriate subject matter? Currently, the most influential voices being heard in this regard come from the efforts of the National Geographic Society, in concert with the National Council for Geographic Education (NCGE), the Association of American Geographers (AAG), and the American Geographical Society. Beginning with the sponsorship of the international survey of geographic literacy in the early 1980s (National Geographic Society, 1988), a concerted effort has been launched to provide the conditions under which, it is believed (or at least hoped), school personnel will see the geographic light. To this end, the AAG and NCGE published *Guidelines for Geographic Education: Elementary Schools* in 1984 (Joint Commission on Geographic Education, 1984). Shortly thereafter, they began organizing the Geographic Education National Implementation Project (known by its acronym, GENIP), which subsequently published two curriculum guides to help implement the *Guidelines* in classroom practice (GENIP Committee on K–6 Geography, 1987, GENIP Committee on 7–12 Geography, 1989). Paralleling these developments, a number of ''Geographic Alliances''—essentially interest groups whose purpose is to promote the study of geography—have been formed at the state level. Their purpose is to help teachers interpret and put into practice the recommendations of the GENIP publications. The National Council for Geographic Education has continued to coordinate the publication of a number of teacher-centered materials designed to enhance the teaching of geography in the classroom.

None of this has been accomplished without controversy. In attempting to resolve issues surrounding the problem of what to teach, there has been an attempt to identify in rather broad strokes what constitutes GEOGRAPHY, writ large, at least for purposes of instruction. Every discipline engages in this sort of self-analysis from time to time. Where geography is concerned, the most noble and extensive treatment of this sort is found in a work titled *Perspective on the Nature of Geography* (Hartshorne, 1959; see also Hartshorne, 1939). Hartshorne was not concerned with geography as it might be taught in elementary or secondary schools; his interest was in explicating it as practiced by professional geographers. However, from his work, numerous other attempts to describe the nature and scope of this thing called ''geography'' have stemmed.

Describing the dimensions of geography one might expect to find being taught in elementary and secondary schools therefore became a matter of interest to those concerned with establishing a more visible geography curriculum. The GENIP guidelines define what might be termed *school geography* as being concerned with five themes. These have generally gone under the rubrics of location, place, relationships within places, movement over the earth, and regions. There has been a general acceptance of these themes demarking ''school geography,'' if the very large number of articles appearing over the recent past in the *Journal of Geography*, the official publication of the National Council for Geographic Education, is any indicator. This development has not arisen without criticism, however. Many academic geographers see it as a perversion of what have been widely accepted, at least at the college level, as the major traditions of their profession, namely: the Spatial tradition, the Area Studies tradition, the Man-Land tradition, and the Earth Science tradition (Harper, 1990; Pattison, 1962, 1990).

The differences in curriculum and teaching implied by adherence to the themes over the traditions may in fact be other than what they appear. As Philip Gersmehl (1992) points out, both *traditions* and *themes* imply a curriculum based on a deductive approach to teaching. That is, teaching is presumed to be based on identifying a subject matter which students are then, hopefully, led to understand; it is, of course, a straightforward approach when we tell a student, "here is the idea, now we'll go through a number of exercises so you will understand it." It is a no-nonsense, efficient way for students when the measure of learning is a correct response in a classroom question-and-answer (QA) session or test. We know, of course, even after information is stored in short-term memory, there is no guarantee it will survive for long. As Gersmehl suggests:

> [F]or some topics, a straightforward **deductive** approach . . . works best: "Here's the theme, and this is how we apply it in the real world." But for many ideas, an inductive approach . . . is more effective: "here's a story, a situation, a problem: let's examine it for awhile; here's another one, a little bit different" . . . The deductive approach usually gives results more quickly, if by "results" we mean correct answers on tests that measure short-term recall of factual material. But the inductive approach can have a much more lasting effect, because it compels the student to apply the tools of analysis and to seek the theme wherever it may be hiding in the welter of everyday experience. (p. 119)

Gersmehl argues that the themes can be most effectively used when they are utilized inductively. That is, the theme serves as a guide for the teacher/curriculum maker, but it is hidden from the direct view of the learner, who will begin to discover its various essences through his or her own inquiry. Indeed, this is a variation on Jerome Bruner's discovery learning (1961).

Teaching based on inductive principles can be an uncertain adventure, and as a consequence, many teachers are not attracted to such an approach. Especially averse to the idea are those teachers who lack an appropriate background for what they are expected to teach. They seek direction and specific forms of guidance, and so the attraction to deductive teaching methods is at least understandable. However, I concur that, when instruction (telling, directing) supersedes education (educing, drawing forth), we do our students no favors, regardless of the subject matter. As Gersmehl points out, "If the five themes are prescribed as the sole remedy for sick geography classes, they will surely become a straitjacket that will eventually limit our ability to devise even more effective ways of teaching about places and place relationships (1992, p. 120).

In what follows, I take the position that a definition of school geography based primarily on the notion of process (or induction) rather than on a product or content (deduction) base will be more helpful in thinking about ways to improve geographic understandings. The "bring geography back" argument is based (precariously, if history is to be our guide) on a definition of school geography which, after all is said and done, is not new. It is in some ways a throwback to a time when *geography* was considered a distinctively separate subject in the school curriculum. Not only is this not in keeping with the realities of academe, it is out of step with curriculum realities. In the first instance, developments in all the social and natural sciences, if not in the entire range of human inquiry, are increasingly demonstrating the artificiality of the traditional subject matter divisions. Although tradition remains a strong deterrent to breaking down the old organizational barriers, even in cases where they persist, the range of academic interests within any given field are broadening dramatically. In precollegiate education, the merging of the traditional separate subjects (history, geography, civics, etc.) was a largely "done deed," in the period following World War I. With the rise of the *social studies*, the natural science side of geography suffered even more than its human ecology aspects. That *geography*, however defined, became the waif of the social studies curriculum also cannot be denied. However, I believe that any reform that brings about a resurgence of interest in geographic inquiry—namely, the study of spatial interactions—will need to be conceived within a unified, rather than a separate, context. In fact, geography is increasingly conceived as an integrated subject rather than a separate one in a number of countries where *geography* in its traditional sense has long occupied a centrol role in the curriculum. Moves in this direction are being reported, for example, in such diverse countries as the United Kingdom (especially England), the former USSR, Australia, and even in the bastion of academic geography, Germany (Marsden, 1990; Maksakovsky, 1990; Gerber, 1990; Schrettenbrunner, 1990).

## REFERENCES

Bestor, Arthur E. 1953. *Educational wastelands: The retreat from learning in our public schools*. Champaign: University of Illinois Press.

Bestor, Arthur E. 1955. *The restoration of learning: A program for redeeming the unfulfilled promise of American education*. New York: Knopf.

Billings, Neal. 1929. *A determination of generalizations basic to the social studies curriculum*. Baltimore, MD: Warwick & York.

Boehm, Richard G., Bierley, John, & Sharma, Martha. January/February 1994. The bete noir of geographic education: Teacher training programs. *Journal of Geography*, 93.1, 21–25.

Boyd, William Lowe. 1979. The changing politics of curriculum policy making for American schools. In Jon Schaffarzick and Gary Sykes, eds., *Value conflicts and curriculum issues: Lessons from research and experience*. Berkeley, CA: McCutchan Publishing Corporation, ch. 3.

Bruner, Jerome. 1961. *The process of education*. Cambridge, MA: Harvard University Press.

Buros, O. K. 1975. *Social studies tests and reviews: A monograph consisting of the social studies sections of the seven mental measurements yearbooks (1938–1972)*. Highland Park, NJ: Gryphon Press.

Caldwell, Otis W., & Courtis, Stuart A. 1925. *Then and now in education 1845–1923: A message of encouragement from the past to the present*. Yonkers-on-Hudson, NY: World Book Company.

Cuban, Larry. 1993. *How teachers taught: Constancy and change in American classrooms, 1890–1990*. New York: Longman.

Department of Supervision and Curriculum Development, National Education Association. 1938/1940/1946. *Building America: Illustrated studies on modern problems*. 9 vols. New York: Americana Corporation.

Eve, Raymond A., Price, Robert, & Counts, Monica. Fall/Spring 1993–1994. International geographic literacy among a sample of U.S. university students. *Phi Delta Kappa International Review*, 4, 1–22.

Flesch, Rudolph F. 1955. *Why Johnny can't read—And what you can do about it*. New York: Harper.

Foshay, Wells. 1990. Textbooks and the curriculum during the Progressive Era: 1930–1950. In *Textbooks and schooling in the United States, 89th Yearbook of the National Society for the Study of Education*, Part 1. Chicago: University of Chicago Press, pp. 23–41.

Geographic Education National Implementation Project Committee on K–6 Geography. 1987. *K–6 Geography: Themes, key ideas, and learning opportunities*. Geographic Education National Implementation Project. Chicago: Rand McNally.

Geographic Education National Implementation Project Committee on 7–12 Geography. 1989. *7–12 Geography: Themes, key ideas, and learning opportunities*. Geographic Education National Implementation Project. Chicago: Rand McNally.

Gerber, Rod. 1990. Geography in Australian education. *GeoJournal*, 20.1, 15–23.

Gersmehl, Philip J. May/June 1992. Themes and counterpoints in geographic education. *Journal of Geography*, 91.3, 119–123.

Goodlad, John I. 1964. *School curriculum reform in the United States*. New York: Fund for the Advancement of Education.

Goodlad, John I. 1984. *A place called school: Prospects for the future*. New York: McGraw-Hill.

Gould, Stephen J. 1981. *The mismeasure of man*. New York: Putnam.

Grobman, Hulda. 1970. *Development curriculum projects: Decision points and processes*. New York: F. E. Peacock.

Hanna, Paul R. 1987. *Assuring quality for the social studies in our schools*. Stanford, CA: Hoover Institution Press.

Hanna, Paul R., Sabaroff, Rose E., Davies, Gorden F., & Farrar, Charles R. 1966. *Geography in the teaching of social studies: Concepts and skills*. Boston: Houghton Mifflin Company.

Harper, Robert. January–February 1990. The new geography: A critique. *Journal of Geography*, 89.1, 27–30.

Hartshorne, Richard. 1939. The nature of geography: A critical survey of current thought in the light of the past. *Annals of the Association of American Geographers*, 29, 173–658.

Hartshorne, Richard. 1959. *Perspective on the nature of geography*. Chicago: Rand McNally.

Hockett, John A. 1927. *A determination of the major social problems in American life*. Contributions to Education No. 281. New York: Columbia University Teachers College.

Holt-Jensen, Arild. 1988. *Geography: History and concepts*. 2nd ed. Totowa, NJ: Barnes & Noble.

James, Preston E., & Martin, Geoffrey J. 1981. *All possible worlds: A history of geographical ideas*. 2nd ed. New York: John Wiley & Sons.

Jensen, Arthur R. 1980. *Bias in testing*. New York: Free Press.

Joint Committee on Geographic Education. 1984. *Guidelines for geographic education:*

*Elementary and secondary schools*. Washington, DC: Association of American Geographers and National Council for Geographic Education.

Karplus, Robert. 1967. *A new look at elementary school science: Science Curriculum Improvement Study*. Chicago: Rand McNally.

Maksakovsky, Vladimir. 1990. The new stage of restructuring of school geography in the USSR. *GeoJournal*, 20.1, 49–53.

Marsden, W. E. 1990. The role of geography in education in England and Wales. *GeoJournal*, 20.1, 25–31.

Murray, Charles, & Herrnstein, Richard J. 1994. *The bell curve: Intelligence and class structure in American life*. New York: Free Press.

National Commission on Excellence in Education. April 1983. *A nation at risk: The imperative for educational reform*. Washington, DC: U.S. Government Printing Office.

National Geographic Society. 1988. *Geography: An international Gallup survey*. Washington, DC: National Geographic Society.

National Society for the Study of Education. 1992. *The social studies in the elementary and secondary school, 22nd Yearbook, Park 2*. Bloomington, IL: Public School Publishing Company.

Olney, J. 1842. *A practical system of modern geography; or, A view of the present state of the world*. 38th ed. New York: Robinson, Pratt & Co.

Pattison, William D. November 1962. High school geography project begins experimental year. *Journal of Geography*, 61, 368.

Pattison, William D. September/October 1990. The four traditions of geography. *Journal of Geography*, 89.5, 202–206.

Rugg, Harold O. 1926. Three decades of mental discipline: Curriculum making via national communities. In Guy Whipple, ed., *Curriculum making: Past and present,* Twenty-sixth Yearbook of the National Society for the Study of Education, Part 1. Bloomington, IL: Public School Publishing Company, pp. 48–49.

Rugg, Harold O. 1931. *An introduction to problems of American culture*. Boston: Ginn & Co.

Saxe, David Warren. 1991. *Social studies in schools: A history of the early years*. Albany: State University of New York Press.

Schaffarzick, Jon. 1979. Federal curriculum reform: A crucible for value conflict. In Jon Schaffarzick and Gary Sykes, eds., *Value conflicts and curriculum issues: Lessons from research and experience*. Berkeley, CA: McCutchan Publishing Corporation, ch. 1.

Schrettenbrunner, Helmut L. 1990. Geography in general education in the Federal Republic of Germany. *GeoJournal*, 20.1, 33–36.

Tanner, Daniel, and Tanner, Laurel N. 1975. *Curriculum development*. 2nd ed. New York: Macmillan.

U.S. Office of Education. National Assessment of Educational Progress. February 1990. *The geography learning of high-school seniors*. Series. Prepared by the Educational Testing Service under a grant from the National Center for Educational Statistics and the National Geographic Society. Washington, DC: Office of Educational Research and Improvement, U.S. Office of Education.

Virginia State Board of Education. 1934. *Tentative course of study for Virginia elementary schools. Vol. 1, Grades I–VIII. Vol. 2, Grades IX–XII*. Richmond, VA: Division of Purchase and Printing.

# 2

# Geography in Historical Perspective

The origins of geography are literally buried in antiquity. Its foundations were first laid over 8,000 years ago, during a time of which we have little direct evidence. There is much to speculate about, but fortunately, recent scientific developments have made the past more accessible. Among them, we might mention advances in underwater exploration and the development of satellites and accompanying advances in photographic abilities. (See, for example: George F. Bass, ed., *A History of Seafaring Based on Underwater Archeology*, 1972.)

## ANCIENT ORIGINS

Although much of what we would like to know about the beginnings of geographic knowledge remain shrouded from direct view, we do know that, at least for the Western world, it derives mainly from the lands adjacent to the Mediterranean Sea. From before 6,000 B.C., civilizations flourished there, particularly in the east. We know, among other things, that the climate of the Mediterranean area was a good deal different than it is today. It was, in fact, a lush land of many rivers and canals, which lent itself to a wide variety of agricultural and other pursuits felicitous to the development of trade, exploration, and military ambitions.

We also know that there was relatively easy access overland to the Red Sea and the Persian Gulf, putting more distant lands in the Middle East, and even India, within reach of military conquerors and traders alike. Consequently, contacts of various kinds doubtless occurred over a large area for a long period of time. In addition, we have learned that ideas from another civilization originating in China were known early on by those living adjacent to the Mediterranean.

In China and the eastern Mediterranean, the need to understand the physical nature of the environment became a palpable fact of life.

As our understanding of these ancient times has expanded, it is becoming increasingly evident that both in China and in the areas surrounding the Mediterranean, knowledge of these portions of the world, at least, was more extensive than previously thought. Even then, from what we know now, it would appear that the wisdom of that day, of which we can make some educated guesses, was not the consequence of sudden discovery but the result of an accretion begun in the mists of time. However, within the time frame for which we have concrete information, we know, for example, that the Chinese were keeping weather records as early as thirteen centuries before the birth of Christ. Surveys of Chinese resources and products are extant from the fifth century B.C., and by the second century B.C., such sophisticated measurements as that of recording the quantity of silt carried by its major rivers were being kept (James & Martin, 1981, p. 56). And as early as 128 B.C., we know the Chinese had extended their explorations sufficiently to have come into contact with the peoples of the Mediterranean. It was from this contact that the concept of locating things on the earth's surface by employing a grid of latitude and longitude derives (as did the arrival in the West of the idea of the decimal system).

The relatively meager artifacts from the early history of the Mediterranean region—primarily in the form of clay tablets and inscriptions on pottery and the like—nonetheless make possible inferences about economic and military matters of the time. For example, we get visions of the kinds of water transport and trade in which people engaged. We also obtain a view of the types of ships used in naval and military encounters. And although much is left to conjecture, it is evident there was a great deal of commercial and military activity for several thousand years within the Mediterranean Sea itself, and in all likelihood beyond.

Clearly, the sea provided the friendliest form of passage that trade and conquest required. While the Mediterranean is not a tranquil sea, its waters have always proven to be more hospitable to movement than the land surrounding it. However, in addition to the fact that the Mediterranean can experience some formidable weather, it is also a very large body of water. Thus, while it is enclosed and protected in a general sense, it is large enough to require in many instances navigation beyond the sight of land. Thus, one of the early mysteries about the existence of geographic knowledge emerges. How could these people, who, as far as we know, lacked accurate instruments for determining where they were in relation to where they had been and where they were going, travel safely and efficiently from one place to another?

Virtually no maps, and certainly no maps useful to a navigator at sea, survive directly from this early period or, for that matter, for more than a millennium following the birth of Christ. A few maps have since been discovered, most notably the world map of Ptolemy, created sometime during the second century A.D. Even then, it would not become known to Western Europeans until the fourteenth century because the Middle Ages brought with them a separation between western Europe and the Arab world, where, despite the destruction wreaked by the Crusades, knowledge of the prior period survived to a certain but limited extent in the universities and libraries established by the Muslims over the centuries.

Yet another interesting fact is that the mathematics necessary for making maps had been created many years before the birth of Christ. The principles of plane and spherical trigonometry had been established, although there is argument whether a particular person accomplished that feat—Hipparchus, some claim—or whether, as in most major discoveries, there was a general accretion of geometrical knowledge over a long period of time, which was perhaps rediscovered at a later date. Nonetheless, these were the tools necessary to render data from a sphere to a flat surface—the basic principle of creating an accurate map.

The period of the ancient civilizations, which came to an end not long before the birth of Christ, was therefore a period in which geographic knowledge developed as a result of pragmatic needs. The demands of trade and the lust for power and control over the world (as it was becoming known to those civilizations that grew up clustered around the shores of the Mediterranean Sea) resulted in the accumulation of information that would later be thought of as geography.

## THE BEGINNINGS OF "GEOGRAPHY"

The next stage in the development of what we know today as geography spans a millenium. It has been referred to as the classical period in the development of geographic thought, extending from, perhaps, 500 B.C. to the fall of Rome in the eighth century A.D. It was the time when more systematic attention began to be given to the nature of things—but primarily physical things as they occurred from place to place over the earth's surface—the beginnings, as it were, of the study of the earth as the home of man.

It is often assumed that people living not long before the birth of Christ held naive and inaccurate views of the world. We know of them now from "maps" that were not much more than imaginative drawings and from the lore about Christopher Columbus, who presumably had to fight the widespread notion that the earth was flat. Americans still seem to grow up thinking that people living during those times believed the world was not just flat but located in some sort of region independent of anything else. While some of these notions apparently did invade human minds, particularly in the Western World, we know quite the contrary to be true, at least among the more educated people of the time, for they had for hundreds of years, and perhaps even centuries, been aware that the earth was spherical and even that it was part of a larger universe. As early as the fourth century B.C., for instance, the hypothesis had been proposed that night and day could be explained in terms of a rotating earth. The length of each day and the extent of the year had been calculated with an extraordinary accuracy. Nor should we forget the Druids, who, in what is now the United Kingdom, constructed such edefices as Stonehenge, which traced the movement of the sun and its consequent relationships to the seasons (see, for example, Hawkins, 1965).

The person generally credited with giving us the word *geography* (''earth-writing'') and one of the first to set out deliberately to solve problems that were essentially geographic in nature was—as far as we know—the Greek astronomer, poet, and head of the great Library of Alexandria, Eratosthenes (c. 273–c. 192 B.C.). It was Eratosthenes who decided to take advantage of several commonly understood facts, coupling them with his knowledge of astronomy and mathematics, to calculate the circumference of the earth (see Figure 2.1).

The first of these facts had to do with a very deep well located near the town of Syene (now Aswan) on the Nile River. It had been observed that at noon of the summer solstice, and only then, the sun would shined directly into the bottom of this well. It was thought that Syene consequently lay on, or very close to, the Tropic of Cancer, the line farthest north on the earth where once each year, the sun's rays are directed straight down (a clear demonstration of knowledge of the earth's sphericity as well as of astronomical concepts about the relationships of the earth to the sun). It was also believed that Syene and Alexandria were separated by about 500 miles. (Distances were not measured in miles in those days, but in what units called *stades*.) Eratosthenes calculated, correctly, that if he measured the angle of the shadow cast by an obelisk in Alexandria, north of Syene, also at noon of the summer solstice, he could employ the geometric principle that when two parallel lines are bisected by another line, the angles thus formed are equal. Thus, if he were to calculate the angle cast by the shadow at the obelisk (which turned out to be 7.2 degrees), then he would know the measure of the angle at the center of the earth: that is, the angle at the obelisk equals the angle at the earth's center ($< O = < C$). By dividing 360, the number of degrees in a circle, by the angle formed by the shadow from the obelisk, and then multiplying by the distance between Alexandria and Syene, the circumference of the earth would become apparent ($50 \times 500$ miles = 25,000 miles).

Eratosthenes' calculation was not far from the 1860 circumference calculation of 24,860 miles, although that figure has more recently been modified to account for the fact the earth is not only somewhat flattened at the poles but is constantly in the process of changing its shape, if only a few miles in any given direction. And of course, some scholars claim that Eratosthenes figured the angle of the obelisk to be 7.5 degrees, making the circumference more nearly 24,000 miles. In any event, he was much closer than some geographers who came later, most notably Ptolemy, who insisted on 18,000 miles. This had detrimental effects more than a thousand years later when Christopher Columbus consequently assumed a much smaller world and, thus, came to believe that India might be accessible by sailing west rather than by traveling eastward, as had been the usual route for three thousand years or more. That Eratosthenes' figure was ultimately proved not quite correct did not result from any serious error, how-

**Figure 2.1**
**Earliest Known Measurement of the Earth's Circumference**

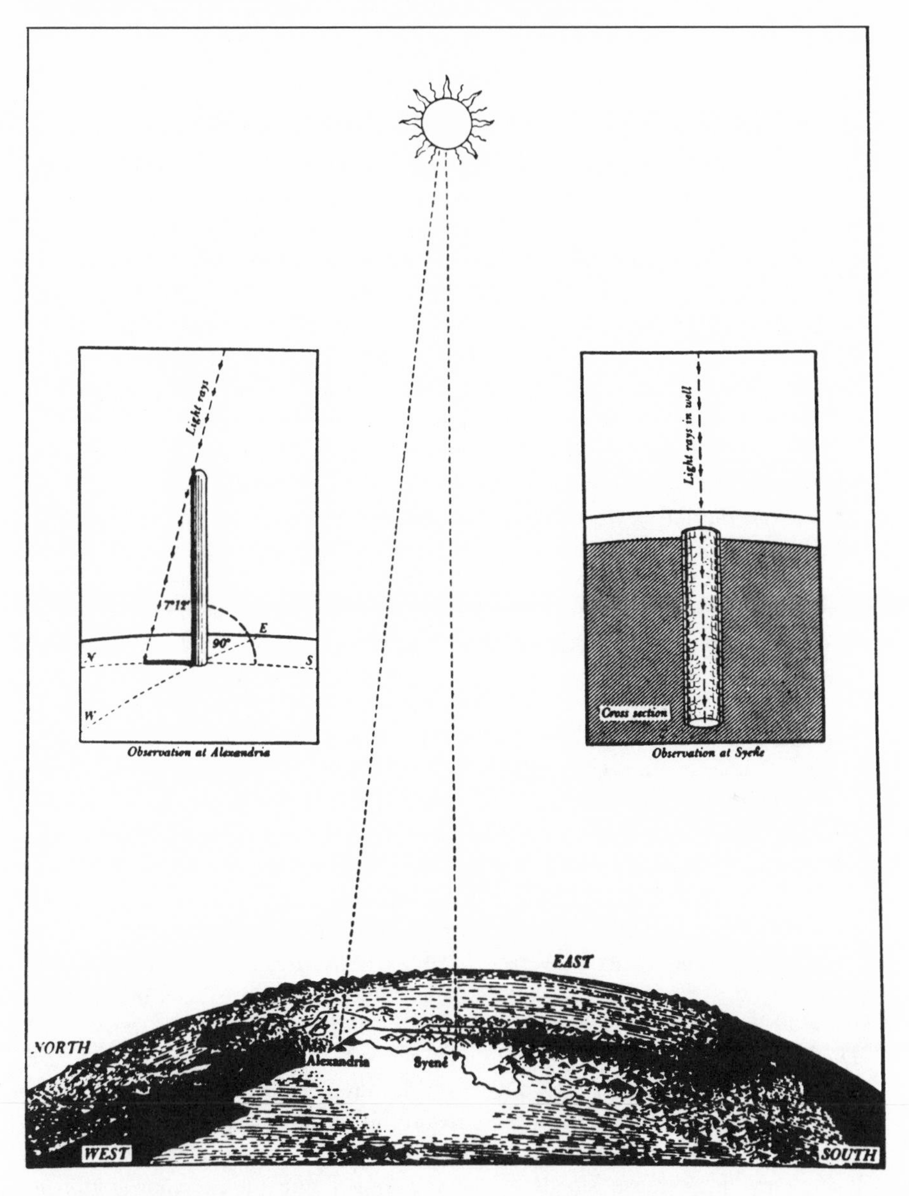


The earliest known measurement of the earth's circumference was made by Eratosthenes about 240 B.C. His calculations were based on (1) the angular height of the sun and (2) the linear distance between Alexandria and Syenê.

*Source*: Brown (1949). Copyright © 1949 by Lloyd Arnold Brown; © renewed 1977 by Florence D. Brown. By permission of Little, Brown and Company.

ever. It was later discovered that the well at Syene and the city of Alexandria were not quite due north and south of each other and thus not on the same meridian; as well, the distance between the two locations was not quite 500 miles. However, these errors tended to cancel each other out.

## GEOGRAPHY LOST, GEOGRAPHY REDISCOVERED

Although many individuals would have collected and recorded geographic information during the Greco-Roman period—extending from about 500 B.C. to the eighth century A.D., little remains from even this period to attest to it. But we do know that during this time, and afterward, despite the destruction brought about by the Crusades, the Arab world became the caretaker of whatever knowledge had been generated during and following the period of the ancient civilizations. While the great library in Alexandria was sacked and burned a number of times, other libraries and universities flourished within the Arab empires from Cordoba, in what is now Spain, to Constantinople in Turkey and beyond.

Many years were to pass before awareness of the existence of this treasure trove of geographic information was to reach Western World. In fact, it was not until about the thirteenth century A.D. that renewed contacts with the Muslim world made possible the translation of these long-lost Greek manuscripts, which the Muslims had translated into Arabic. They were now translated once again, into the common language of written communication in Christian Europe, Latin. This geographical information helped launch the Age of Exploration, which began early in the fifteenth century with Portugal's Prince Henry the Navigator, whose ships explored the coast and islands of western Africa, and then spread outward with the voyages of Columbus, Ferdinand Magellan, James Cook, and a host of other explorer/conquerors.

Finally, in the 1400s, after a silence of some 800 years, two major geographic works became available to the Europeans. One was the seventeen-volume tome titled *Geographica*, written (or perhaps compiled is a better word) by Strabo (c. 64 B.C.–A.D. 20). Strabo had sought to bring together all that was known about the world at the time of the Roman Empire. Originally intended as a sourcebook for statesmen and military leaders, Strabo's *Geographica* was apparently not widely read, for few other scholars of his time referred to this work. But its value in later years proved inestimable because of its service as a repository of much of the geographical information known at the time of its writing—at least in Europe and the Middle East, for it is well to remember that the Chinese were making their own discoveries in this regard, many far ahead of what would be relied upon to advance the Age of Exploration.

Ptolemy, in contrast, stands as the towering figure for his contribution to mapping based on mathematical principles. We know little about the person, but we do know that he, also, worked at the library in Alexandria (between A.D. 127 and 150). Although he accepted as accurate another calculation of the circumference of the earth, the figure favored by his teacher, Marinus of Tyre (which placed it at 18,000 miles instead of 25,000), and although this led to serious miscalculations by Columbus and others, his contribution to the developing of maps using the grid system, which is still in use today, was of enormous importance (see Figure 2.2). Like Aristotle, Ptolemy accepted the opinion that the earth was a sphere, but a stationary one around which the celestial bodies orbited (an idea that would have to await Nicolaus Copernicus to be finally dislodged). Like Strabo's *Geographica*, Ptolemy's *Guide to Geography* is a compilation of the work of others. However, two of its eight volumes contained his explications of the application of mathematical principals to the construction of maps. Like many other map constructions of his time, however, the problem of calculating longitude and expressing it on a flat surface posed a serious difficulty since there was no way at that time for east-west distances to be calculated with any accuracy (a dilemma that waited for solution until the eighteenth century). And while he contributed more than any other to understanding how features on the earth might be represented on a flat surface, his actual maps were also monuments to inaccuracy, resulting primarily from the inability to calculate longitude with any precision. That problem was compounded by Ptolemy's error in accepting his teacher's calculation of the earth's circumference rather than that of Eratosthenes. Together, these errors were to wreak havoc once the Age of Exploration arrived in western Europe.

We have direct knowledge of Strabo and Ptolemy because their work has come down to us in almost complete form, but other geographers of the Greco-Roman period we know only from inference since no artifacts of their work have yet been discovered. What we do know is that, following the decline of the Roman Empire in the seventh and eighth centuries A.D., the secular concerns of Christian Europe fell into oblivion. Now they were those of the Catholic Church, that of fighting off barbarian invaders and ''paganism'' alike. And although the development of geographical understandings effectively came to a halt during the period from about A.D. 700 to the 1400s (a period sometimes called the Dark Ages), it is also well to remember that at the same time the Vikings were exploring the islands of the North Atlantic, including Greenland, as well as the northern shores of what would later become known as the American continent (see, for example, Kirsten A. Seaver, *The Frozen Echo: Greenland and the Exploration of North America, ca. A.D. 1000–1500*, 1996). At the same time, the Muslims were actively expanding the Arab world, eventually influencing an area extending from the Iberian Peninsula (they governed most of Spain and Portugal for over 900 years and would have dominated a good part of western Europe had they not been defeated at the Battle of Tours in 732) to Baghdad (Iraq) and beyond. Evidence also exists of the extensive development of geographical understandings in China. Thus, all was not stagnation as it was in western Europe.

Perhaps the most important set of data about the world as it was known in the 1400s comes to us through the maps that seafaring people used, and apparently had used for some time previously. These are known as *portolan* maps.

**Figure 2.2**
**Sixteenth-Century Ptolemaic Map**

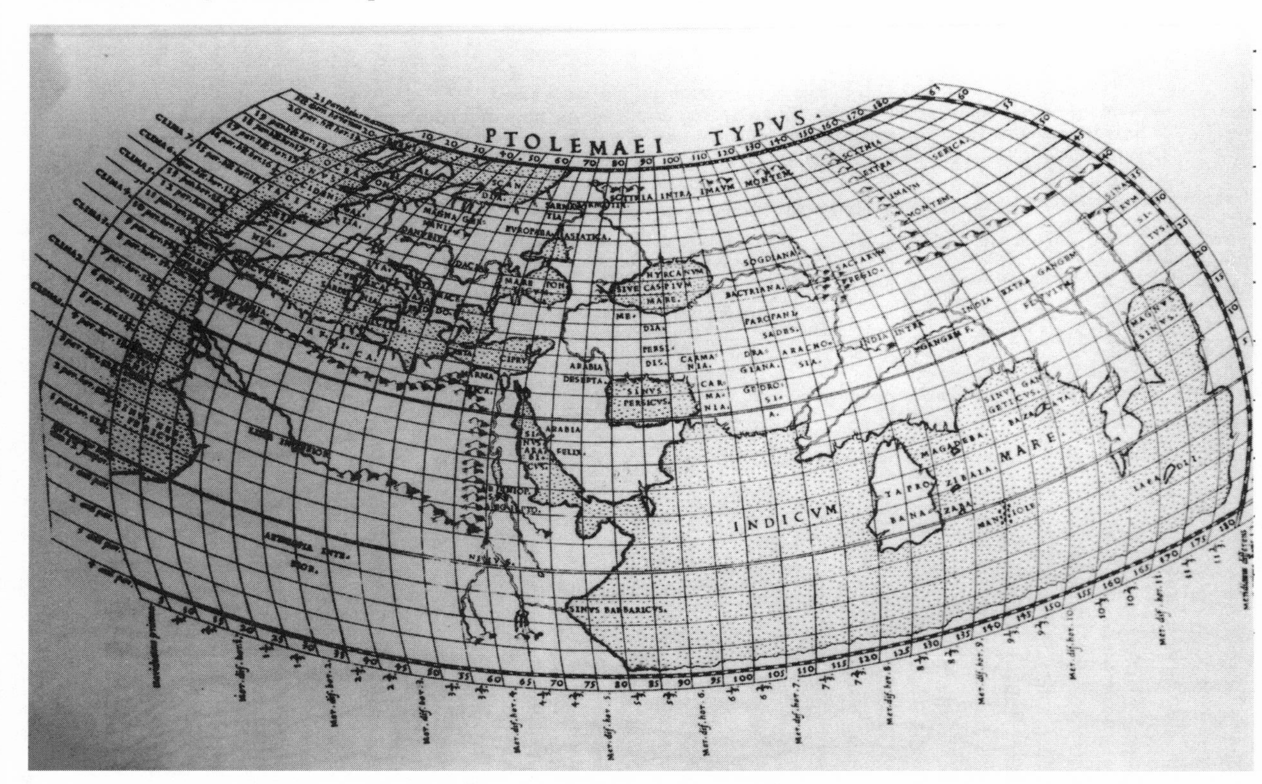


A sixteenth-century Ptolemaic map based on data from his *Guide to Geography*. Note that the ‘‘known world,’’ to Ptolemy and to others for many years thereafter, meant that virtually no detail below the equator was provided. Longitudes are expressed in fractions of hours east of the Fortunate Islands, while latitudes are designated by the number of hours in the longest day of the year.

Figure 2.3 is a reproduction of one from the 1500s. Portolan maps were not widely known in western Europe until the 1500s (the earliest extant one so far discovered dates from the early 1400s), but they clearly could not have sprung up in a matter of only a few hundred years. Basically, the portolan map was a representation of a given area, perhaps even the world, with a series of rhumb lines (a rhumb line represents an easily followed, constant-direction course between two points) indicating the directions of the compass. (That these lines might simply be devices allowing the sailor to set a compass course to the destination begs a larger question of why the centers of these systems were located where they were, which we shall examine shortly.)

The making of maps in those days was a tedious process because the printing press had yet to be invented (or at least known, since there is also evidence that such a device had made its appearance in China prior to that of Gutenberg). Each map was one-of-a-kind, created usually on animal skin, and was a treasured and rare possession of the sea captain owning it.

Since we have not yet discovered any portolan maps dating prior to the fourteenth century, the maps we do know of are doubtless the descendants of many previous efforts, all of which were results of the ongoing experiences of sailors, collected during their travels, the result of compilation, and in all likelihood made of compilation upon compilation. When people set out to create maps, therefore, they doubtless copied portions of a previously existing map and added any new information about the location and direction of places that might have come to hand. And, depending on the extent of the area the mapmakers wished to depict, they might copy anywhere from one preexisting map (in this case, the portolan map) to many.

The portolan (seafarers) maps demonstrate that, long previous to the early stages of the Age of Exploration, which would begin in the late 1500s, humans had quested into the farthest reaches of the earth. Perhaps the most notable discovery in this regard has been the finding of the Piri Re'is Map in the old Imperial Palace in Constantinople in 1929. Originally drawn in 1513 (the Moslem year 919), it was signed with the name of Piri Ibn Haji Memmed, who had been an admiral (''Re'is'') in the Turkish navy—hence *Piri Re'is*. In its original form the map was a rendering of the whole world. Only a fragment has survived the ravages of time and military conquests, however (see Figure 2.4). It has brought special attention, not just because of its early date, but because it is clearly the earliest known map of America. In addition, it also shows Africa and even Antarctica, in what appear to be amazingly correct latitudes and longitudes. This was just a few short years after the voyages of Columbus and long before anyone in Christian Europe knew of even the existence of these places, let alone their relative configurations and locations in relation to one another. What, one might wonder, had the original map revealed about the world?

The Piri Re'is Map, while unknown in Christian Europe, was surely a singular part of the resources of the Turkish admiral who was its author, and consequently, of the navy he commanded. Doubtless there were other maps of its

**Figure 2.3**
**Sixteenth-Century Portolan Chart**

This sixteenth-century portolan chart adds to the confusion by being drawn ''upside down.'' Note the rhumb lines intersecting the coastline at designated locations. How the points of intersection were determined remains a mystery.

*Source*: Nordenskiold (1889).

**Figure 2.4**
**Fragment from the Piri Re'is Map**

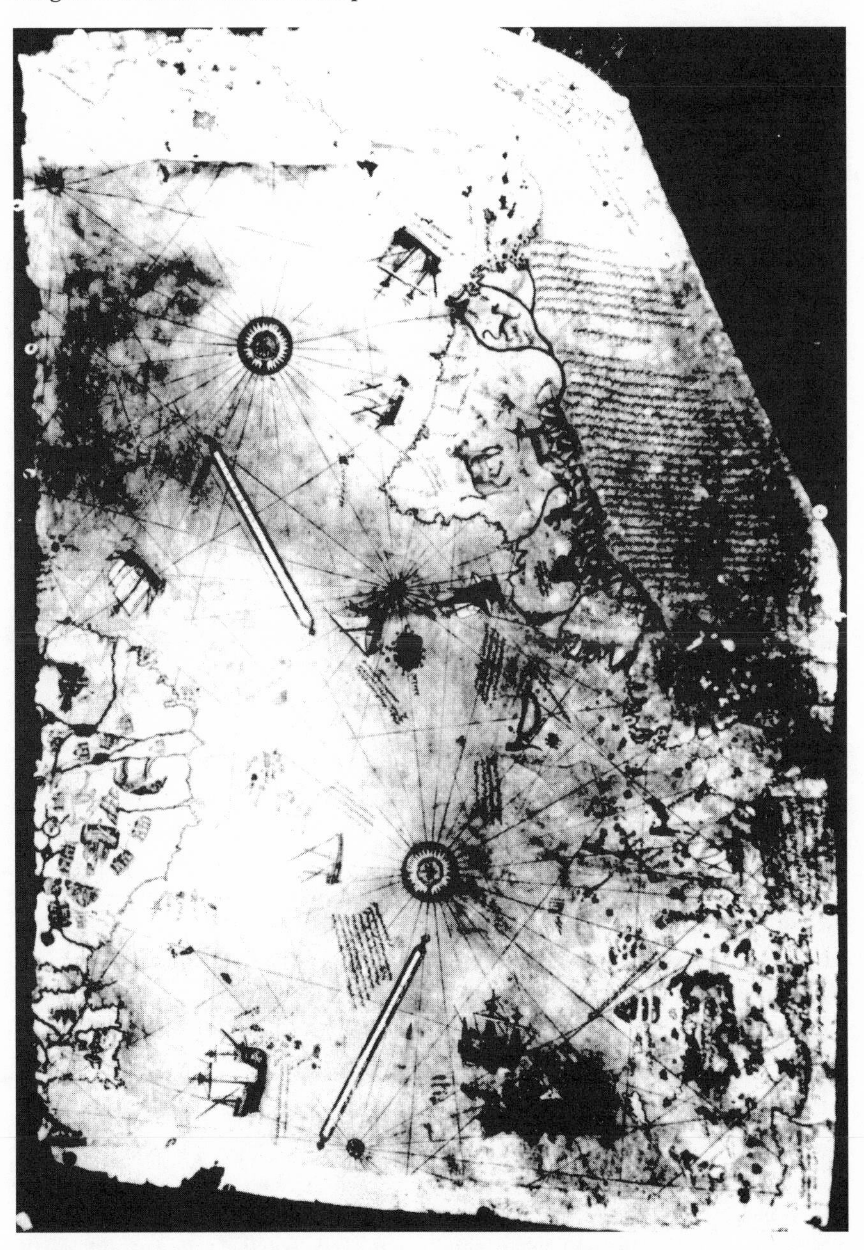

A fragment from the Piri Re'is portolan map shows lands surrounding the Atlantic Ocean. On the right side are the Iberian Peninsula and Gibraltar; to the south is a portion of Africa. On the left are South and Central America and islands of the Caribbean Sea, and to the north are the western coast of North America and Greenland.

*Source*: Topkapi Palace Museum. Reproduced with permission.

sort, rich in detail and revealing lands and oceans acquired through explorations long prior to the "discoveries" that were to take place during the Age of Exploration in the sixteenth and seventeenth centuries. That there were civilizations existing far in advance of what we have previously thought may not now seem such an unlikely idea. This is what Charles Pellegrino suggests in his book, *Unearthing Atlantis: An Archaeological Odyssey* (1991), which relates the story of an expedition that sought to understand the origins of the fabled story of Atlantis, the island which, Plato tells us, disappeared along with its advanced civilization in some ancient time. Such speculation is encouraged because nothing of any certitude is known of human occupancy prior to the establishment of Babylonia around 6000 B.C.

Charles Hapgood has written a book based on data drawn from the Piri Re'is Map itself. In *Maps of the Ancient Sea Kings: Evidence of Advanced Civilization in the Ice Age* (1979), Hapgood asserts that this map, and probably other maps of the period, were compiled from information secured during the period beginning after the last ice age, some 10,000 years ago. He speculates that knowledge compiled during those early years "was incorporated in ancient books, . . . handed down from age to age, and survived in manuscripts Eratosthenes saw in the Library of Alexandria" (p. 179). In any event, we are led to conclude that the explorers/conquerors of the fifteenth, sixteenth, and even seventeenth centuries would have been better off with portolan maps rather than the highly inaccurate work of Ptolemy.

Newer methods of exploration, particularly the advancement of technology that allows the recovery of artifacts from ancient seabeds, will doubtless add greatly to our store of knowledge about this part of the past, which has long been hidden from view. From these artifacts we will learn more about the kinds of ships, and even the tools of navigation, that made possible these earlier geographic discoveries. Ship carpentry and boat-building, for example, are perhaps the most conventional of all trades, with secrets of construction having traditionally been passed down through the generations. Underwater exploration, which is just now coming into its own, will doubtless, not only tell us much about the kinds of ships ancient mariners sailed, but analyses of their cargoes will contain much additional information rich in detail.

## PRECURSORS OF MODERN GEOGRAPHY

The Age of Exploration extended over a period of almost four centuries. Its beginnings date from the early 1400s, when Prince Henry the Navigator first dispatched his Portuguese ships to explore the African coast and the islands lying nearby. It was a time when Ptolemy's map was thought to be a true rendition of the earth and when sailors believed that the region south of the equator, the place Ptolemy called *terra australis incognita* (the unknown southern land), would cause them, at the very least, to turn black, like the people who were known to live there, as they proceeded down that coast. The age's demise is sometimes associated with the voyages of Captain Cook in the mid- to late 1700s to the Pacific Ocean area. Of course, exploration of the earth's surface, and particularly of its landforms and its geology, has continued to the present time. But the initial stages of its discovery, at least in Western eyes, is often said to encompass this period.

The Age of Exploration was aided and abetted by the refinement of the compass and the compass rose, two instruments basic to navigation, and by the invention of the chronometer which, in its turn, made possible the accurate measurement of longitude.

### The Invention of the Compass and the Compass Rose

The first instrument essential to navigation is, of course, the compass. History fails to tell us when the principle of the compass first became known or how it came to be employed in navigation. Strangely, the earliest documention of the compass as we know it today is in association with the compass rose, which indicates directions (north, east, south, and west, with intermediate directions set within a circle of 360 degrees), and comes to us from the twelfth and thirteenth centuries. Clearly, however, awareness of direction and the means for determining it were known long before that time. For example, if we can accept the possibility that portolan maps existed long before their actual discovery in western Europe, as seems certain, then we can see that compass directions had become part of the technique of navigation from, perhaps, the earliest times. Ptolemy, of course, incorporated the idea of the compass rose when he drew in the figures surrounding his map to indicate the points from which the winds might come (see Figure 2.2).

As well, we know that the characteristics of magnetite, or loadstone, were known very early on. Both Plato and Euripides wrote about its qualities as early as the fifth century B.C. Reference to it is also to be found in early Chinese writings. For example, the following definition appears in a Chinese dictionary completed in A.D. 121: "a stone with which an attraction can be given to a needle" (Brown, 1949, p. 126). Although the discovery of the characteristics of magnetite remains shrouded in the past, we can be certain that the Chinese were not the only ones to discover that a needle might be similarly magnetized and that when it was attached to something that would float, that it would point northward, in the general direction of the polestar. Such a discovery, crude as it might be, was essential to sailors who would maintain their course, as it freed them from the necessity of having cloudless nights to determine the direction of their sailing.

Compasses showing deviations from north in terms of degrees of declination from magnetic north had thus become a genuine instrument of navigation perhaps a hundred years before Prince Henry's time. By then, compasses had been designed that balanced a magnetized needle on a stationary pivot. One Petrus Perigrinus became the first to assign a position to the poles of the loadstone, proving that a magnetized needle possessed both positive and negative properties and that unlike poles attract. As important, he was the first (in the Western world, at least) to surround the magnetic needle, so mounted, within a graduated circle. Soon after, the compass rose, heretofore long associated with wind directions (as in Ptolemy's world map), appeared with the magnetized needle. Thus there came into being the navigational compass we know today. There were yet to be refinements in manufacture but, more important, refinements also occurred in our understanding of the differing effects of the magnetic fields of the earth. Although the phenomenon of the magnetic declination was not understood until relatively recent times, its effects quickly became evident as sailors continued to find themselves off their intended course.

Knowledge of direction is so fundamental to the practicing navigator, as well as the map or chart maker, that we sometimes fail to realize how significant the discovery of the qualities of the loadstone were or how it influenced the development of the science of navigation and the maps and charts that came into being as a consequence.

Early seafarers needed, not only to know direction in the very practical sense of guiding their ships from one place to another; they required maps or charts of their destinations to be as accurate as possible. After all, one would hope to know the location of reefs and shoals, of islands, and of ports of call beforehand if a successful landfall were to be made. And while we are not certain when the principle of the magnetic compass first came into use, we do know that the refinement of this instrument, as in the case of the development of more accurate tools to measure both latitude and longitude, was absolutely essential to the success of the Age of Exploration.

## The Discovery of Latitude

The principles on which was founded the concept of latitude—the measurement of distance north and south of the equator—were known long before the Age of Exploration. Eratosthenes drew on them when he calculated the circumference of the earth. In that calculation, he used the tool that is basic to all measurements of latitude, the *gnomon*. In his case, the tower in Constantinople served this purpose, although a gnomon, technically speaking, is any rod or pin utilized to cast a shadow that shows the position of the sun. In its earliest form, the gnomon was nothing more than a tree or a stick that, by casting a shadow on a sunny day, told the reader how the day was progressing. It later became used to indicate the passage of the seasons and, in fact, was the first tool that allowed observers, in noting the changes in the ratio between the height of the gnomon and its shadow over the seasons, to begin the process of identifying the parallels of latitude on the map.

Eratosthenes' tower obviously could not be moved as a rod might, but then again, even that kind of gnomon would not do its work on the deck of a ship bobbing about on the ocean. History consequently records a search for an instrument that was sufficiently portable to carry anywhere a person might find it worthwhile to know the latitude. The problem of calculating latitude, while known in ancient times, therefore did not lie with its mathematics, but with the lack of portability of the gnomon.

One of the first such refinements of the gnomon was an instrument known as the astrolabe. We are uncertain when the first crude astrolabe came into existence, but it surely must have been hundreds of years before the birth of Christ. In its simplest form, the astrolabe consisted of a quadrant marked off in degrees, affixed to which was a movable arm. With that in hand, the operator could measure any angle. For the navigator, it made possible the measurement of the height of the sun from the horizon, or at night, the measurement of the polestar or any other object in the sky. Knowledge of that angle, when combined with charts indicating the position of any one these objects at a particular time of year, made possible the calculation of latitude with considerable accuracy. We know the astrolabe in its more modern dress—which evolved during the early 1700s—as the sextant. The astrolabe/sextant, however, did have two serious failings. Rough seas could reduce its accuracy considerably. Worse still, if the sun did not shine at noon or if the weather did not clear sufficiently to see the stars at night, it was useless. Today, of course, all such calculations are made electronically. Technology has now advanced so rapidly that all the necessary electronics needed for establishing one's location on the earth's surface can be contained within a hand held instrument, the accuracy of which now provides the answer within a foot or so. We know this now as the Global Positioning System (G.P.S), an instrument available for a mere one hundred dollars or a little more in its simplest form.

### The Invention of the Chronometer and the Calculation of Longitude

The idea that the earth's surface might be mapped by laying out a system of lines in the pattern of a grid had been thought out by the Chinese long before Ptolemy's time. Just how the location of particular places within that system might be fixed was, of course, quite another matter. We also know that scholars were able to calculate latitude long before Eratosthenes used his formidable mathematical skills in estimating the size of the earth. And the charts of the declination of the sun for each degree of latitude had obviously also existed for a considerable period prior to the achievement of his remarkable feat.

Longitude was another matter, however. It was not until the Age of Exploration was nearly over that the techniques and equipment necessary to accurately calculate distances east and west, in degrees of longitude, were finally devised. The delineation of longitude—the calculation of distances east and west—was fundamental to the construction of accurate maps, and consequently to the development of contacts involving great distances, whether they be military in nature or for the purposes of trade and immigration. Movement over the face of the earth, and especially at sea, remained chancey until a solution to this problem was found. Again, it was not that the principle of calculating longitude was unknown; rather, the difficulty lay in the development of tools by which it might be accomplished. Until that day arrived, sailors contented themselves with guessing how fast their ship had traveled in a day; many believed that longitude at sea was a measurement that would never be made with mathematical accuracy. But the vicissitudes of the times, notably the conflicts that were breaking out between Spain and Portugal over the territories the Pope would make accessible to each for purposes of exploration (and plunder), required accurate demarcation.

Much as was the case with latitude, the first mathematical measurements of longitude were made on terra firma. But this required more than the knowledge of mathematical formulas. It was preconditioned as well by two other developments: the invention of a clock of considerable accuracy, and the ability to construct a telescope of sufficient power to observe (as it turns out) the moons of Jupiter. However, it was the need for a chronometer—a clock that could keep time with exceptional accuracy—that turned out to be the most critical element in the equation, and the most difficult of solution.

Galileo (1564–1642) provided the first set of observations necessary to the much-sought calculation. In his observations of the universe and his consequent confirmation of the Copernican theory (that the earth was part of a solar system rotating about the sun and not itself the center of the universe—for which he was excommunicated and not formally forgiven until recently), Galileo had discovered that Jupiter was surrounded by a number of moons, which orbited in an identifiable pattern. With a telescope powerful enough to observe the moons clearly, Galileo was able to time the exact moment at which an eclipse of Jupiter by one of the moons would begin and end. Moving his equipment to another location, he could determine the difference in the beginning and ending of an eclipse between the two locations, and from those data, he was able to establish the longitudinal distance between the two.

The accurate determination of longitude awaited one remaining fact: the calculation of the earth's diameter. Eratosthenes' figures had long since been abandoned because of the influence Ptolemy still held over the scientific community. A new effort to establish the exact distance between each degree of longitude was then undertaken in France, where King Louis XIV had brought together many of the most respected scientific figures from all Europe. The group adapted methods from Eratosthenes and others, and the resulting figure of 7,801 miles was remarkably close to the one accepted today, and the one originally calculated by Eratosthenes (Brown, 1949, p. 218), from which it became possible to demark the circumference into 360 degrees with a high degree of accuracy.

Determining longitude in the real world at this point required a telescope mounted on a firm base, clearly a condition that could not be transferred to a ship at sea. Without a dead calm (a relatively rare event), it would be impossible to get a fix on Jupiter's moons. Moreover, since clocks of the day were regulated by pendulums, they were totally unsuited to life aboard ship. But a combination of an accurate chronometer and a gnomon/sextant would provide the necessary condition for a reasonably accurate reading of longitude. Thus, there was launched an effort to encourage the development of a chronometer that could be taken to sea—an effort that would take many years before success would be achieved. In 1714, therefore, the English Parliament passed a bill "for providing a publick reward for such person or persons as shall discover the Longitude" (Brown, 1949, p. 227), and the search was on. It would not even begin to become a reality until 1735, as Dava Sobel recounts in rich detail in her book, *Longitude: The True Story of a Lone Genius Who Solved the Greatest Scientific Problem of His Time* (1995). That year, John Harrison, a self-educated Yorkshire carpenter, invented and built (following seven years of labor) the first marine chronometer, which weighed in at 72 pounds. There followed three more versions, each lighter and more accurate, succeeded by sea trials at which the standard of accuracy set by the Board of Longitude, three seconds a day, was more than met. Harrison's first successful clock was completed in 1728, and the fourth in 1759.

The successful construction of a chronometer meant that the determination of longitude no longer depended on sighting Jupiter's moons through a telescope. Rather, a simple noontime fix on the sun's angle in the sky was all that was necessary. The process of determining longitude remained the same, however. That is, an accurate chronometer made possible the determination of differences in observations of the angle of the sun from one place to another over time (with the forerunner of the modern-day sextant), the calculation of which gave an accurate reading of the distance between the two points at which the readings took place.

Although the Age of Exploration (1400–1700) was fast coming to a close when Harrison's chronometer (and its successors) came into use, its appearance marked the first time highly accurate maps could finally be made. As new information flowed back to Europe from the many explorers (who were bankrolled, not only by their sovereigns, but as well by business interests eager to tap the resources of these "newfound" lands), the need for more accurate maps became self-evident. But that was not the only impetus for improving knowledge of the location of things. At home, kings and queens realized that they required credible records of the nature and extent of their realms at home as well as abroad, not simply in dealing with foreign affairs, but for domestic purposes as well. The cartographic profession came into its own as a result of all these converging factors and one other: the earlier appearance of the printing press. Johannes Gensfleisch, or Gutenberg (1400?–1468), was reputedly the first in Europe to successfully print with movable type. By the sixteenth century, and thereafter, advances in the technology of printing elevated the reproduction of maps to new levels of importance wherever the flag of the realm might be flown.

## THE ROOTS OF MODERN GEOGRAPHY

The so-called classical period in geographic thinking began to give way to modern geography, its study as we know it today, a relatively short time ago. The demise of the classical approaches and the founding of modern geography are frequently linked to the deaths, in the same year, 1859, of the two most famous geographers of the nineteenth century, Alexander von Humboldt (1769–1859) and Carl Ritter (1779–1859). One might also note the publication that year of Charles Darwin's *Origin of the Species.*

However, it would be a mistake to assume that these three events marked a sudden shift from the familiar to the new. In that regard, recall the legacy of earlier years, particularly the teleological views that permeated western European thinking, including the eighteenth-century ''construct of the Great Chain of Being that arranged all of creation on a vertical scale from animals and plants and insects through humans to the angels and God himself'' (Gates, 1992, p. 55), which by extension was then, appallingly, said to account for a hierarchy contrived to validate the superiority of the white race. Teleology comprises the belief of a divine hand in natural occurrences, the conviction that natural phenomena are determined by an overall design or purpose beyond that derived from direct observation. Ritter, particularly, ''saw in all of his geographical studies the evidence of God's plan (James & Martin, 1981, p. 129). Thus did learned men of their day account for ''natural'' calamities as well as the spectacular and usual events observed from place to place. Just as the earthquakes, typhoons, water spouts, and optical phenomena such as the Aurora Borealis were believed to be divine manifestations of God's hand, so were the differences in human characteristics and the opportunities and tragedies making up the human condition. We have not divested our thinking of such ideas, of course; many persons, particularly those of a conservative religious persuasion, still hold to the notion of a divine spirit determining individual fates and conditions.

Geography as a subject to be taught had entered the curriculum of the universities prior to Humbolt's and Ritter's tenures. The most notable practitioner of such teaching was Immanuel Kant (1724–1804), who is generally thought of today as a philosopher (see May, 1970), but someone who also taught ''geography'' for the better part of his career. It was Kant, in fact, who first defined geography in more modern terms, as ''a narration of occurrences which are coexistent in space'' (quoted in James & Martin, 1981, p. 323). Humboldt and Ritter were to be the first persons whose teaching focused exclusively on the subject of ''geography,'' and from their students would come a next, and larger, generation of people who called themselves geographers.

One of the next generation was Arnold Guyot, Ritter's student. He is important to the development of geography in the United States for his scholarly contributions and his extraordinarily influential school textbooks and, in both contexts, for his adherence to Ritter's teleology and the Great Chain of Being (see, e.g., Guyot, 1885). Baccalaureate education along with geographic instruction in the public schools was consequently influenced early on by the notion that all natural phenomena were in some ways also manifestations of a divine hand. In school textbooks, teleological implications were only slightly hidden by descriptions of major catastrophes—earthquakes, for which the causes were yet to be known, are said to have mysterious origins and connections; waterspouts, hurricanes, and the like appear as both random physical occurrences and divinely guided phenomena, thus becoming especially dramatic representations evidencing that power in all nature (see Figure 2.5). One can only wonder at the extraordinarily prejudicial information contained in the textbooks of the day.

The discoveries of Charles Darwin proved to be the harbinger of today's natural and social sciences (see, e.g., Darwin, 1859). The decades immediately preceding the turn of the century were witness, particularly, to an intense interest in studying human behavior modeled on the forms of scientific inquiry that were increasingly coming to characterize research in the natural sciences. This was a period during which there was a very rapid expansion of higher education in the United States and, along with it, the establishment of the academic disciplines we know today as the social sciences: economics, political science, sociology, psychology. It was a period of both discovery and specialization. Within this frame, the two fields of synthesis, history and geography, were to take quite different paths.The study of history found a special niche in a country with a very short history of its own, but with roots in the western European tradition. It consequently commanded instant recognition as the source for cataloging, categorizing, and in fact giving meaning to the development of a distinctly American tradition. Geography, in contrast, found itself torn between its past as inquiry into the nature of the physical world and the new demands of developing empirical generalizations and even perhaps hypotheses and theories about spatial phenomena. While the history of the world might be described within a chronological frame, it might also be so spoken of within a spatial context. But the delimitation of history proved to be no thorny problem; scholars selected from it those contexts within which the nationalistic and cultural resources might find explanation, namely to western Europe and the origins of its Judeo-Christian ethic. Geography was not to be so fortunate. There were simply not available the exclusionary criteria that history enjoyed. For geography, the whole world was quite literally its oyster; but there were no compelling criteria that would make one kind of geography more significant for study than another, as was the case for history.

Even more difficult a problem was that brought about by the invention of the social sciences themselves. The question arose as to whether geography was a physical science, a social science, or both. Heretofore primarily concerned with physical phenomena, in which human beings were largely at the mercy of natural (and divine) forces, the idea of the earth as the home of humanity (where human activity might take advantage of, as well as being subject to, its control), introduced a new element in considering the value of geography and its

**Figure 2.5**
**Illustration Taken from *Cornell's Physical Geography*, a Popular Text of the Late 1800s**

EARTHQUAKE IN CALABRIA, IN 1783.

An illustration from a popular geographic text of the late 1800s shows the teleological emphasis permeating not only the natural world, as in depictions of natural disasters, but also its heavy-handed assumptions regarding the bases of human behavior.

**Figure 2.6**
**The Circumference of Geography**

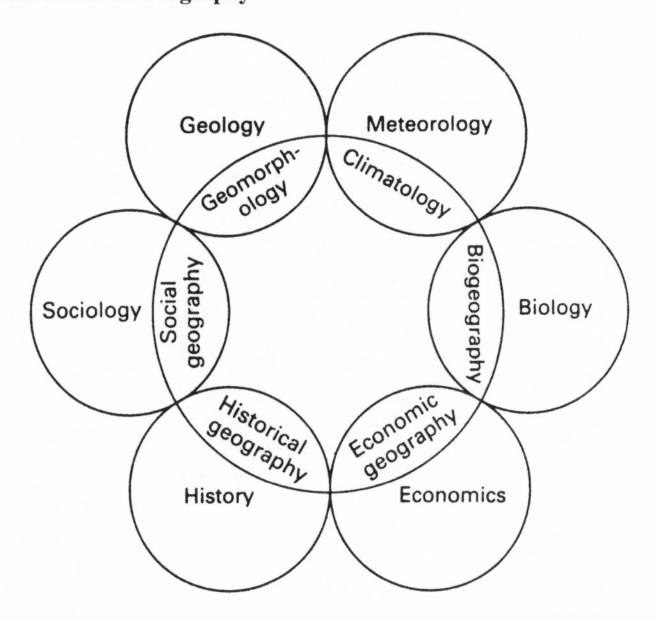


*Source*: Adapted from Fenneman (1919); Holt-Jensen (1988). Used with the permission of Barnes & Noble.

utility as a field of study. At first, and in some instances continuing to this day, the study of geography as an area of academic interest found itself housed within the rubric of geology. It was not what one might think of as a robust subject or topic for academic study. By 1914, for example, at the advent of World War I, courses in geography were being offered in only fifteen universities throughout the United States. Figure 2.6 depicts the wide-ranging and overlapping nature of the concerns which geographic inquiry potentially encompasses.

As higher education expanded, there was an understandable interest on the part of the new social scientific community to seek to clarify the parameters of their disciplines, geographers among them. Although the teleological notions of the Ritters and Humboldts had been fairly well laid aside, a new form of determinism had crept into the scene. The new determinism derived, curiously enough, from Darwin's theory of evolution. Although it is doubtful that geographers living after the turn of the century read very much of Darwin, the belief became widespread that science would provide specific answers to explain human behavior. The possibility that behavior is predetermined, with its rules only wanting discovery, had an infinite attractiveness. Geographers were not the only ones to be so smitten. We see such an attraction, for example, in the undeniably important work (even today), of Jean Piaget, who began in the 1920s to postulate a theory of an invariant cognitive developmental pattern. Neo-Piagetian revisions of the more deterministic aspects of his work are now increasingly forthcoming (see, for example, Donaldson, 1978).

So persistent was the problem of determinism in geographic thought (as in the social sciences generally), that a hundred years after Darwin, the issue was still very much alive. For example, Richard Hartshorne, the distinguished geographer of the second and third quarters of the twentieth century, wrote:

> Many scientists who recognize that science cannot demonstrate the hypothesis of scientific determinism nevertheless adhere to it as an article of philosophical faith to be defended as the foundation on which the whole structure of science depends. Any suggestions of doubt, any assumption of the possibility of free will, must therefore be attacked with wrath and scorn as unscientific. (1959, p. 154)

Forty more years have now passed, and we are not entirely out of the woods. While geographers generally have rid their profession of determinism, I mention it here because it remains in the minds of nongeographers, including those who would teach geographic ideas. On reflection, it would appear self-evident the existence of certain environmental features, or the particular location of an area in relation to others, does not of necessity dictate human behavior. It can all too easily be demonstrated, for example, that humans living in essentially the same physical environments in different parts of the world have historically responded to the opportunities and restrictions of these places in fundamentally different ways. Yet the notion lives on (particularly where geography and politics intersect), the negative aspects of which are best illustrated in the case of Nazi Germany between the two world wars.

World War II brought an unprecedented demand for geographers. It was not just the GIs who were ignorant of geographic facts in the places where they would be called on to fight, but the generals as well. A lack of accurate maps, climatalogical knowledge, and almost everything else, at least in sufficient detail to be useful, plagued the government. The 300 geographers recruited to fill the knowledge gap was hardly enough match for the problem. Crash courses to train military and civilian personnel were instituted. Geographic literacy had become a major public policy issue. However, courses in geography were still more likely than not to be included within other divisions of the university—again, mainly geology—under the rubric of earth sciences. For those not smitten with a love for geography, including many students being instructed at the time, the poor reputation of geography as an interesting subject for study unfortunately became further reinforced.

Once the immediate need was past, geography once again suffered a major identity crisis. Rather than offerings expanding in American colleges and universities, quite the opposite happened. Several major institutions closed their departments (for example, Stanford and Harvard universities), while many others never did get around to offering work in geography except inasmuch as they

**Table 2.1**
**The Establishment of Geography Departments in American Colleges and Universities, by Year**

| YEAR | Number of Distinct Depts. Of Geography | Geography Combined With Other Fields |
|---|---|---|
| Before 1990 | 1 | 3 |
| 1900-1909 | 8 | 3 |
| 1910-1919 | 7 | 2 |
| 1920-1929 | 15 | 6 |
| 1930-1939 | 8 | 9 |
| 1940-1949 | 29 | 4 |
| 1950-1959 | 17 | 13 |
| 1960-1969 | 33 | 30 |
| 1970-1979 | 8 | 10 |
| 1980-1989 | 2 | 1 |
| Since 1990 | 2 | 3 |
| Totals | 130 | 84 |

*Source*: Association of American Geographers (1996).

participated in the war effort. Indeed, the names of institutions that have never had a geography department read like a who's who list of higher education. Among them we find Princeton, Yale, and all the other ''Ivy League'' schools. Neither the Massachusetts nor California institutes of technology has had one (Dorschner & Marten, 1990). And in the face of a rapidly expanding overall national population, geography as a field of academic interest has declined, and the number of professional geographers with it. Tables 2.1 and 2.2 illustrate this strange predicament.

Table 2.2 shows that the number of degrees conferred has been stable over a considerable period of time, despite rapid growth in college enrollments and the population generally. The high ratio of male to female students completing degrees is also notable.

The past quarter-century, particularly, has been a period of dramatic technical developments affecting how we are able to observe earth phenomena: new means of mapping with aerial and satellite technology, the development of radar, infrared, and satellite imagery; more sophisticated mathematical concepts and statistical procedures; the advent of the electronic computer; and a myriad of other exotic developments. Once again, geography as a distinct field of study has been endangered by the development of numerous subfields requiring extensive specialization. It appears to be a case of déjà vu for the ''mother of sciences,'' which for example, initially gave rise to the classical subfields of geology, chemistry, economics, and so forth.

**Table 2.2**
**Geography Degrees Conferred in the United States, 1947–1956 to 1993–1994**

| | BA/BS | | MA/MS | | Ph.D. | |
|---|---|---|---|---|---|---|
| | M | F | M | F | M | F |
| 1947-1956 | 4482 | 1148 | 1281 | 303 | 328 | 26 |
| 1957-1966 | 9490 | 1889 | 2114 | 379 | 536 | 50 |
| 1967-1976 | 28448 | 8039 | 5412 | 1254 | 1523 | 112 |
| 1977-1986 | 23331 | 10412 | 4184 | 1751 | 1069 | 275 |
| 1987-1994 | 19149 | 8877 | 3175 | 1684 | 784 | 281 |

*Source*: Association of American Geographers (1996).

## GEOGRAPHY TODAY

All scholarly fields engage in self-analysis. Each asks, ''What is the nature of our field?'' ''What is English?'' is a frequently asked question, strange as that may seem, and ''What is Geography?'' also has occupied the attention of the geographic profession for a very long time. Indeed, of all the scholarly professions, geography has perhaps been the most torn by this always devisive question. Hartshorne's *Perspective on the Nature of Geography* (1959), has been looked on as the most definitive statement of the scope and nature of geographic inquiry. But the argument lingers, in major part because, as other fields have continued to spring from ''geography,'' however described, the need to define has continued, if only because there seemed always to be less and less of the field left for scholarly inquiry as it is increasingly being split into subfields. What defines geography's uniqueness?

Although many have attempted to describe its singularity, there seems to be general agreement that the primary focus of geographic inquiry ought to be upon spatial interactions (Edward L. Ullman, *Geography as Spatial Interaction*, 1980). That is, how can areal phenomena best be understood? How can we explain the interactions between the human characteristics and physical conditions as they are arranged over the earth's surface? What can we learn from the understandings that derive from such study?

The problem of the modern geographic scholar, then, is to identify the spatial or areal conditions within which spatial interactions can best be analyzed. Illustrating the long-standing inability to agree on what orientation the study of geography should take is an ongoing argument. The idea of regions—analyzing interactions within a region, however defined—has been one of the major points of departure in modern geographic inquiry. Other approaches have included studies based on interactions of specific phenomena utilizing an area larger than that of the region per se. Such studies are topical in nature and may set out to answer questions that are, for example, economic, agricultural, historical, or religious. Simply put, any problem or question can be studied geographically, that is, from a spatial interaction point of view. Thus we see geographic study that inquires into the distribution of natural resources, of agricultural activities, and of disease. We may also find analyses of various kinds of migrations, of the nature and extent of different kinds of trade, and so forth. It is not the *nature* of the question that defines geography but the *method* of inquiry.

The geographer makes meanings about spatial interaction through the process of mapping, that is, by recording information within a system of grids representing the area selected for study and, subsequently, by drawing inferences from the data entered thereon. While maps provide a singular means for preserving spatial information, it has, until relatively recently (if we cast our net to include the total existence of the idea of *geography*, or earth writing), drawn upon art, religious conviction, and even fantasy as much as or more than on science. The basic scientific information needed to describe location with the necessary precision has been a relatively recent development—one encompassing hardly more than 200 years. Then, it should be realized that the truly sophisticated tools on which modern geographers draw—aerial photogrammetry techniques from satellites, the electronic computer, and so forth—have come into widespread use only since the 1960s.

## MEANINGS FOR TEACHING

In Chapters 1 and 2 I have reviewed the history of geographic instruction in the United States, summarized past and current dismay over the effectiveness of that instruction, and traced the emergence of geography as an academic discipline. What sense can be made of this account where teaching and learning are concerned?

First, it is evident that the kind of information widely said to indicate knowledge of geography is, at best, a superficial indicator of geographic literacy. Second, it appears instead that "geography" knows no particular subject matter. Rather than an emphasis on the accumulation of facts and study focused upon a specific set of phenomena, it is concerned with spatial interactions of many diverse kinds. We can think about virtually anything in a geographical way, much as we might think of such things in the context of history. Third, it therefore seems obvious that the primary problem of the school in the development of geographic concepts is one of finding ways to help students think spatially—to think in terms of spatial interactions.

Spatial awareness is fundamental to the generation of meanings about spatial interactions. How does the student acquire such awareness? How does one learn to make and read a map, the tool geographers of all ages use to record the data from which spatial interactions are both intuited and deduced?

Surely, one needs "facts" in order to establish relationships. But it is not enough to stop with the discovery of facts; understanding relationships—spatial interaction—requires emphasis be focused on that aspect of learning. Lucy Sprague Mitchell made the same point in a remarkable book, *Young Geographers* (written in 1934), when she asked:

> Do children learn by a passive accumulation of facts or by putting two facts together and discovering a new fact, which is the relationship between them? Which in a general way, is asking whether children learn more by thinking or by being trained? How do they arrive at a process called thinking? What, for that matter, takes place when *any one* thinks? Is not thinking essentially the seeing of relationships? Until an individual through some active process puts 2 and 2 together and discovers 4 we do not feel that he is thinking. The 4 is discovery, almost his creation. Whether it is a simple direct sensory relation such as "fire burns," or a complex relation in which a common element is abstracted from many diverse experiences, such as "people who live in congested cities depend upon distant works for food, etc., upon transportation facilities, etc." The process in either case is active, dynamic, and creative. It is utterly different from parrot-like repetition—the giving back of facts just as they are found. It is *using* facts; not merely knowing them. (p. 10)

As Mitchell also points out, the question is not confined to young children but to students (or learners) at any age.

I shall suggest in the following pages that geographic concepts derive from *doing* geography. That is, one moves toward geographic literacy by exploring the environment, at first through direct experiencing, and, as cognition matures, by more vicarious or abstract avenues. Always, the focus is upon understanding relationships—upon developing an appreciation for spatial interaction.

Such understanding is, however, dependent upon developing the ability to interpret and read maps. I take the position that the acquisition of this ability is very much like learning to read print, except that it is in fact a much more complex process. Increasingly, it is being agreed that one learns to read print much more effectively when extensive writing precedes, or at least accompanies, experience with the printed word (e.g., Goodman & Goodman, 1979; Douglass, 1989; Willinsky, 1990). Similarly, geographic literacy appears to depend on the ability to create one's own map. Quite literally, then, I believe that map writing should preceed, or at the very least accompany, learning to read a map. These are details that will be explored in the chapters that follow.

First, there is a caveat that comprises an exceptionally important codicil to the problem of teaching and learning where geography is concerned. It is generally conceded that it is difficult, if not impossible, for teachers to teach effectively if their experiential/knowledge base is meager or lacking in important respects. By implication at least, from the data I have reported here, which is detailed and amplified elsewhere (Bluet, 1981; Association of American Geographers, 1993; Dorschner & Marten, 1990, James & Martin, 1981), we can see how difficult it is to bring the ideas unique to geography to the teaching profession. There simply are not enough informed people, relative to the need, to accomplish the task, at least if we think of the problem of teacher education in a traditional sense.

In other words, teachers are no more likely to respond to directions about how to teach (directions that are themselves derived deductively) than are their students. Like their students, they need to "mess around" with geographic ideas if they are to perceive geographic thinking as not only something more than, but as different from, knowledge compiled from sets of disparate facts. They need to realize that *geography* is in fact the process of thinking about phenomena in a spatial context in which maps serve as means to an end, and not as a repository of isolated bits of information.

Textbooks, workbooks, and other materials that purport to provide the basic framework for the curriculum fill a "third party" role in the teaching-learning dyad. While many teachers prefer to share their responsibility for teaching with a textbook, many others do not, and it is in these classrooms that we are most likely to see innovation and high levels of student interest. There is, of course, a greater inclination to rely on materials prepared far beyond the classroom door (more often than not by "authorities" whose credentials are unknown), when the topic or subject matter is particularly unfamiliar or when pressures mount to raise achievement levels. When these forces combine, their effect is particularly detrimental to teacher creativity and, I would suggest, to learning generally.

Unfortunately, people who set out to change or "improve" curricular practices, whether they are of a professional or lay stripe, tend to short-circuit the situation by preparing materials that are deliberately intended as a substitute for teacher knowledge. Rather than providing knowledge for teachers about the nature of geographic thinking, various kinds of lessons are contrived that are designed to be directly applicable in the classroom. Where electrical energy is concerned, we all understand what results when a short-circuit occurs; its analogy in the classroom appears to produce a much more benign but just as significant result. That is, we quickly appreciate that a problem exists when the lights go out, but we do not respond commensurately when the light inside one's head fails to go on.

To make the act of teaching something that belongs to the teacher, *not* to some distant (physically as well as intellectually) authority, a preceptor model other than a substitute for the teacher's brain is needed. One that I believe holds particular promise is found in the National Writing Project. (See, for example, Robert L. Root & Michael Steinberg, eds., *Those Who Do, Can: Teachers Writing, Writers Teaching: A Sourcebook*, 1996.) The basic assumption underlying this now-extensive effort to improve the writing ability of secondary school students particularly is that, in order to teach writing, one must have experience as a writer. Amazingly, upon investigation, we find that many teachers have never had *any* extensive writing experience. They are products of our schools, of course, yet for some reason (which may not be too difficult to divine), their teachers failed to provide them with a background in writing: they may have studied a lot, but they did not write. To provide this missing ''writing experience,'' groups of teachers participating in one of the National Writing Projects join together under the leadership of someone who has an understanding of the writing process to learn about the process itself, not in some didactic way, but by engaging in writing for its own sake. The premise, now widely accepted as valid, is that teachers who find their own ''voice'' as writers in the process acquire the kinds of abilities that teach their own students to write. In this process, the teacher abandons the role of an intermediate transfer agent of information or skills that some distant ''expert'' has arbitrarily determined to be significant for the role of participant. That there are parallel possibilities in, for example, developing geographical thinking abilities has yet to be explored. Yet it seems to provide a promising model for enhancing geographical teaching abilities. In many ''in-service'' efforts there seems to be the assumption teachers must be provided with knowledge about methods of teaching lest they founder in figuring out what to do in an actual classroom. My assumption is that when one feels comfortable with ideas, ways of bringing those ideas to life in the classroom follow rather naturally. Teachers are a variable bunch, and so what emerges in classrooms also will vary, but there are many approaches to teaching, and so to demand that teachers follow a particular pattern or sequence is neither reasonable nor productive in changing behavior.

But here the focus is on helping already experienced teachers. What should we do about teachers-to-be? There is general agreement about the inadequacy of today's typical teacher education programs. Preparing for teaching in perhaps 95 percent of our colleges and universities today involves four to five years in which students combine an academic program with a series of courses designed to prepare them for teaching. While the trend has been toward extending the period of ''training'' (we call it that although one might hope it is a misnomer and does not actually describe what is going on), the general pattern of courses and their content have changed little over the years. Typically, students begin by taking ''foundations'' courses which deal, albeit superficially, with the historical and philosophical roots of American education. Students then move on to a series of methods courses. The sequence is capped by an extended period of ''student teaching.'' The methods courses, particularly, are almost universally derided for being simple-minded and are frequently referred to with terms such as ''kiddie-math'' or ''kiddie-lit.'' In the final phase, the teachers-to-be are assigned to a ''master teacher,'' with an emphasis on the ''real world'' of the classroom. The classroom is commonly viewed, by both mentor and mentee, as the place where the future teacher finally learns to teach—not surprisingly, in the image of the master teacher, who conveys (in one way or another but usually with great effectiveness), that it is now time to discard the silly notions students have presumably learned in their college classes. There is no doubt of the powerful influence of the supervising or master teacher who confirms the irrelevancy of theory while passing the mantle of traditional classroom behaviors to the next generation.

This is an unfortunate situation. One consequence is that untold numbers of highly qualified people shy away from the mindlessness of traditional teacher education programs (although not a few find their way into teaching after trying other professional opportunities). Another is that even the people who stick it out tend to deride their experience; it is not unusual to hear these teachers-to-be say that the time they must spend in the teacher education sequence is not unlike serving a prison sentence. Nor is it any wonder that whole new generations of teachers arise who have little time for theorizing about or contemplating how they might improve their teaching.

There *are* alternatives, primarily those based on the concept of an internship in which theory and practice are intermixed over an extended period, which includes the responsibility of full-time teaching. Under these circumstances, the student/teacher plays both roles simultaneously; in this circumstance, the "real world" of the classroom finds utility in theory as a guide to action.

Tradition is so strong in American colleges and universities that only some sort of crisis seems to be able to dislodge it, and then only temporarily, where initial teacher preparation programs are concerned. When there have been teacher shortages, "experimentation" with convention has, to a certain extent, been allowed; but when the shortage abates, the new approach is abandoned for the usual way of doing business. This has, I submit, a chilling factor on bringing about change or innovation in our schools. Given the already existing shortage of geographic "expertise," added to which are the traditional curriculum and the resistence to change (as well as legislative strictures imposed in licensing requirements), the situation looks difficult, if not bleak. So-called preservice teacher education programs will probably not change much if those who teach in them do not change.

## REFERENCES

Association of American Geographers (AAG). 1996. *Guide to programs of geography in the United States and Canada, 1996–1997*. Washington, DC: AAG.

Bass, George F., ed. 1972. *A history of seafaring based on underwater archeology*. New York: Walker & Company.

Bluet, Brian W., ed. 1981. *The origins of academic geography in the United States*. Hamden, CT: Shoestring Press.

Brown, Lloyd A. 1949. *The story of maps*. Boston: Little, Brown & Company.

Cornell, S. S. 1876. *Cornell's physical geography*. New York: D. Appleton & Company.

Darwin, Charles A. 1859. *Origin of the species by means of natural selection; or, The preservation of favored races in the struggle for life*. New York: A. L. Burt Company, Publishers.

Dilke, O. A. W. 1985. *Greek and Roman maps*. Ithaca, NY: Cornell University Press.

Donaldson, Margaret. 1978. *Children's minds*. New York: Norton.

Dorschner, Donald L., & Marten, Robert O. May/June 1990. The spatial evolution of academic geography in the United States. *Journal of Geography*, 89.3, 101–108.

Douglass, M. P. 1989. *Learning to read: The quest for meaning*. New York: Teachers College Press.

Gates, Henry Louis, Jr. 1992. *Loose canons: Notes on the culture wars*. New York: Oxford University Press.

Goodman, K. S., & Goodman, Y. M. 1979. Learning to read is natural. In L. B. Resnick & P. A. Weaver, eds., *Theory and practice of early reading*. Hillsdale, NJ: Lawrence Erlbaum, 1:137–154.

Guyot, Arnold. 1885. *Physical geography*. Rev. ed. New York: American Book Company.

Hapgood, Charles H. 1979. *Maps of the ancient sea kings: Evidence of advanced civilizations in the Ice Age*. Rev. ed. New York: E. P. Dutton.

Hartshorne, Richard. 1959. *Perspective on the nature of geography*. Chicago: Rand McNally.

Hawkins, Gerald S. 1965. *Stonehenge decoded*. Garden City, NY: Doubleday.

Holt-Jensen, Arild. 1988. *Geography: History and concepts*. 2nd ed. Totowa, NJ: Barnes & Noble.

James, Preston E., & Martin, Geoffrey J. 1981. *All possible worlds: A history of geographical ideas*. 2nd ed. New York: John Wiley & Sons.

Landstrom, Bjorn. 1961. *The ship: An illustrated history*. Garden City, NY: Doubleday & Company.

May, J. A. 1970. *Kant's concept of geography and its relation to recent geographical thought*. Toronto, Canada: University of Toronto Press.

Mitchell, Lucy Sprague. 1934. *Young geographers*. New York: John Day Company.

Morrill, Robert W. June 1991. One perspective on geography. *Perspective*, 19.5, 59.

Nordenskiold, A. E. *Facsimile Atlas of the Early History of Geography*. Trans. J. A. Ekelof & C. R. Markham. Stockholm, 1889.

Pelligrino, Charles. 1991. *Unearthing Atlantis: An archaeological odyssey*. New York: Random House.

Root, Robert L., & Steinberg, Michael, eds. 1996. *Those who do, can: Teachers writing, writers teaching: A sourcebook*. Urbana, IL: National Council of Teachers of English; Berkeley, CA: National Writing Project, University of California.

Seaver, Kirsten A. 1996. *The frozen echo: Greenland and the exploration of North America, ca.* A.D. *1000–1500*. Stanford, CA: Stanford University Press.

Sobel, Dava. 1995. *Longitude: The true story of a lone genius who solved the greatest scientific problem of his time*. New York: Walker & Company.

Ullman, Edward L. 1980. *Geography as spatial interaction*. Seattle: University of Washington Press.

Willinsky, J. 1990. *The new literacy: Redefining reading and writing in the schools*. New York: Routledge.

# 3

# The Development of Spatial Concepts

**Grade Five Geography Lesson**

Children never get to the point.
They surround it.
The importance of the point
Is the landscape of it.
You begin by discussing
"The Rainfall of Vancouver Island"
And somebody has an uncle who
lives there.
And there is an uncle in Alberta
Who has a zillion cows,
Some chickens and a horse
(We get to feed the chickens
and ride the horse),
Which brings us to an uncle
In Saskatchewan, who has a house
where
Deer pass the kitchen window
Every morning (he takes us out
And shows us where they go).
If there were no uncles on
Vancouver Island
It would never rain there.

Barry Stevens

## A WORD IN GENERAL

Thinking geographically means, beyond everything else, exercising a capacity that everyone possesses to one degree or another: the conjuring of mental images about the spatial environment. The ability to imagine, or otherwise construct in one's mind, percepts about space is a learned capability, the consequence of direct experiencing *and* intellectual stimulation, but also of development itself. In the Stevens poem, which is written from a Canadian perspective, we are reminded of how important first-hand experiencing is to a child who is asked to visualize things or places beyond his or her immediate perception. Although such discussions are seemingly always fated to get off track in classroom talk, they are in fact, very much on-target; such "digressions" are nothing more than a way of seeking relevance for something that in reality is quite abstract.

In this and the following chapters the reader will find a discussion of a number of aspects relevant to the development of spatial understanding. In the present discussion, the development of various aspects of the individual's concept of space itself are explored. In Chapter 4, I discuss the process of map "reading," or how purposeful attention to the instruments the cartographer creates to both record and to represent spatial interactions within a given physical arena become meaningful to the reader.

It is generally thought that there are perhaps three distinctive aspects to spatial thinking; we can at least isolate them for the purposes of discussion and research, even though they are not ultimately isolable, one from the other, in actual life. First is the idea of *perspective*: how one views the relationships, one to the other, of those elements that we might perceive in an environment. Growth in this ability implies the capability of being able to take the Copernican view, namely, *to "see" the relationship of things in the environment from something other than one's own personal viewpoint.*

A second idea has to do with notions of *location*, namely, that things have a particular place within the environment and that distance and direction act independently to give each a unique identity relative to the environment as a whole. The third idea deals with concepts of *elevation*. Not only do phenomena occupy a particular space, which has both mathematical and relative relationships with other environmental elements, they exist on a vertical as well as a horizontal scale. Interconnecting them is what the geographer calls "relief," namely, the nature of the slope on which each of the environmental elements in a given area is situated.

The ideas of perspective, location, and elevation are, as one might assume, full of complexities. Even in their less mature forms, they possess a deceptive simplicity, encouraging instruction that is not age appropriate. As formal instruction is increasingly being urged for younger and younger children, we should consider carefully what we know about intellectual development. Basically, we are informed by two major sources. One is the research of Jean Piaget (1896–

1980) and his colleagues, which is recorded primarily in four books published in English between 1954 and 1970 (Piaget 1954, 1968; Piaget & Inhelder, [1948] 1956; Piaget, Inhelder, & Szeminska, 1960). The second comes to us through a number of researchers who, stimulated mainly by Piaget's pioneering work, produced a spate of studies in the 1960s and 1970s designed primarily to replicate or extend his findings. Since the early 1980s, however, there has been only an unsteady trickle of such studies. Might this reflect a shifting of interest toward the accumulation of knowledge versus the process by which knowledge is generated?

Although the data are not as extensive as one might like, important insights remain to be gleaned, but within a context begging to be understood where Piaget's work is concerned. By this I mean certain ambiguities in his constructs describing the growth and development of children's cognitive abilities. We may understand how important his contributions to our understandings of spatial concepts have been but must also recognize the points at which some care should be taken in interpreting his ideas.

Piaget has been called the giant of the nursery, referring to his amazingly creative research into the development of children's thinking, which he conceptualized and conducted over a period of some 60 years. It is, however, important to appreciate why his research methodology was for many years dismissed as a serious contribution to our understanding of human growth and development. For although Piaget's research reports began to be published in Europe in the late 1920s, his ideas were effectively kept from the mainstream of American psychological thought until the seventh and eighth decades of the current century. As a Swiss, he published his books in French, but they were soon thereafter translated into English. His ideas became widely known early on in England and to a lesser extent in Great Britain generally. However, it took until 1950 before one of his books, *The Psychology of Intelligence*, would be produced by an American publishing house. It would take another decade, and more, before his other writings began to become available to American scholars. As a consequence, his influence in American education came late and remains still to be widely understood and appreciated, particularly at the practitioner level in our elementary and secondary schools.

Piaget's methods conflicted with mainstream research in the United States. The major point of contention lay in his decision to study intensively a very few children at any one time, sometimes as few as two or three, rather than employing a very large sample. The latter approach has been, and remains, the usual procedure in the United States, where the test items—usually in paper and pencil form—cover a much wider range of topics. Piaget's purpose in adopting his procedure, which he called the *clinical method*, was based on a major difference he had with his American counterparts. Rather than being interested in a child's ability to give a correct response to questions, Piaget sought to understand the reasoning processes that led children to give incorrect answers to the tasks he set before them to solve. As well, instead of attempting to secure answers in a third-person setting—the paper and pencil test—he asked youngsters to talk their way through the solution to a particular kind of problem.

From his findings, Piaget developed a theory of intellectual development at odds with traditional notions, where growth is seen as a steady, quantitative progression to the mature state. Rather, he argued, the use of intelligence follows a qualitative series of changes, moving along an irreversible course through four major stages and their several substages, along with the ages in which each type of thinking dominates. The first of the four major shifts in the quality of a child's thought he denoted the *Sensorimotor Period*; extending from birth to about two years, it is characterized at first by purely reflex activity, followed by hand-mouth coordination via the sucking reflex, which then evolves into hand-eye coordination and the establishment of object permanence (understanding, for example, a person's face still exists even if he or she covers it from its view). The final stages of this period find the infant able to internalize representations and to combine them in what amounts to making new connections between familiar objects. It is, for example, a time when the meanings of ''yes'' and ''no'' become established and when ''mine,'' but perhaps not ''yours,'' emerges. It is the period of growth that is most dramatic as far as the quality of change in the development of the thinking process. It is, however, also the period of infancy least amenable to direct teaching, although this may also be said to a very large degree for the second stage, termed the *Preoperational Period*.

This period extends from the age of about two to seven years. This is the time during which the child gains mastery over the syntax and grammar of the particular language community. It is when, for example, the double negative in the form of words such as ''ain't'' and ''isn't'' becomes part of the working language and impervious to being dislodged by mere instruction, namely, *telling* the youngster what is acceptable and what is not. It is the time when vocabulary burgeons, when the order in which words are arranged in the mother tongue is settled, and when the grammar of that language is acquired. Research in language development now clearly shows that these developments also occur as a result of an inner capacity for finding order, and not simply because parents, much as they may have tried, have sought to instruct the child in such matters. More important for our purposes, it is also a time when intellectual problems cannot be solved logically, before acquisition of what Piaget called the ability to ''conserve,'' that is, to use logic to solve problems. Children at this stage of development find solutions to problems through their own perceptual apparatus, not by the application of logical reasoning. Solutions to problems are reached on the basis of what they seem to be, most frequently through use of the visual process. In the Piagetian examples that follow, we see that the child at this stage finds solutions to spatial problems by concluding what they must be on the basis of the problem's appearance.

The third period has been termed the *Concrete Operational*; it commences, according to Piagetian theory, around the age of seven and extends to about eleven years. (These age boundaries have been questioned by subsequent researchers, most of whom now agree that the age span for this and the other periods has a great deal more variation than Piaget perceived.) It is during this period in the growth and development of children's thought that ''reversibility'' is achieved, meaning that thought is no longer dominated by perception, and that as a consequence the child is able to use logic apart from what is perceived with the senses to solve problems. There is a limitation, however: logical mental processes can be applied successfully in situations where concrete problems are present, but in the case of complex verbal problems without reference to concrete objects, the child's reasoning powers fail.

Finally, then, there is the expectation that the last barrier to complex thought, namely the inability to reason hypothetically, will drop by the wayside. Piaget expected this to occur sometime between the ages of eleven and fifteen years. However, this period as been the least studied of the four, and there appear to be situations at either end of the age scale where this does not hold true.

Piaget did argue, however, that in all the areas in which reasoning might be classified (e.g., spatial, mathematical) and for which subclassifications might be designated, development invariably occurred along what might be called a broken front (what Piaget called a *décalage*), where the child handled one reasoning problem in one period, or stage within a period, or even between periods.

Piaget's view is not entirely unlike the overly deterministic nature of geographic thinking during the late nineteenth and early twentieth centuries. In Piaget's conception, intellectual development is seen as an inexorable march through the first three stages and their various substages to the fourth stage, which also has special aspects. We can excuse Piaget's rather rigid assumptions about development since he was, after all, the product of a time when the idea of stage development (actually an outgrowth of social Darwinism), became popular. Piaget's determinism was, of course, not his exclusive idea in psychological circles, any more than were the ideas of Ritter and Humboldt. Perhaps the best example from a near contemporary were the ideas held by an extraordinarily influential American psychologist at Yale University, Arnold Gesell, who categorized features of human growth and development in much the same fashion but with an emphasis on social and physical aspects (see, for example, Gesell, 1930, 1940; Gesell & Ilg, 1946; Gesell, Ilg, & Bates, 1956).

We are increasingly becoming aware of the excessive rigidity in both Piaget's and Gesell's work, as well, as in the work of other writers, on aspects of human growth and development during the first two-thirds or so of the twentieth century. We are, for example, becoming much more aware of cultural differences and their potential effect on the kinds of problems different societies necessarily address. Piaget's concept of formal operations, for example, is defined very largely within the framework of western European traditions. If his ideas regarding this aspect of cognitive development were applied in, say, Senegal or rural areas of Mexico or other Latin American countries, or among the Marshall Islanders, who developed maps unlike any we have seen in the Western world as sophisticated navigational guides, the evidence suggests that many fewer citizens of those places would meet Piaget's requirements for hypothetical thinking. There are, doubtless, several brands of such thinking, which we have yet to discover (Bruner et al., 1956). See, for example, Figure 3.1, which shows a Micronesian sailing chart.

Awareness is growing, therefore, of certain limitations to Piaget's "clinical method." Studies are beginning to show how both the research design and the verbal interactions that Piaget believed revealed children's thought processes might produce unnecessarily rigid results. The best example comes to us from a description Margaret Donaldson included in her little book, *Children's Minds* (1979). In a replication of Piaget's famous Mountains Task (described in detail in this chapter), the experiment was deliberately modified to make the experimental situation more accessible to children's thinking. The results showed much younger children demonstrating a wider range in reasoning competence than Piaget had postulated (Donaldson, 1978).

Even with caveats such as these, there is no question that Piaget has led us to new ground in understanding the emergence of thinking in children and young adults. His most enduring legacy has been to point out that intellectual development is a *qualitative* rather than a *quantitative* phenomenon. And while we may expect modifications in Piagetian theory to continue, his picture of human growth and development has not only changed our views of growth and learning in radical ways, it has been given much greater detail through his efforts and by those who have replicated and expanded on his work.

## A WORD ABOUT WORDS

One of Piaget's major contributions to understanding the development of thought processes has been to reduce our dependence on words as indicators of children's understandings. Most of us tend to take what children tell us to be honestly reflective of what they know. However, Piaget realized how deceptive this tendency can be and devised experiments that put a major reliance on demonstration. Because children, particularly from the age of 18–24 months onward, grow so rapidly in their ability to speak, it is very easy to be led to believe that understanding follows along at a relatively equal pace. This is particularly the case during the preschool period, when the average child amasses a speaking vocabulary of some 12,000 words (an acquisition rate, on average, of 8 or 9 words every day). This rate of expansion continues at least throughout the school experience, although it is not strictly associated with school attendance as vocabulary acquisition appears to continue during the summer as well as the other months of the year. Estimations of vocabulary acquisition between the third and twelfth grades show a continuation of this growth curve averaging 3,000 words per year—a continuance of the average daily acquisition rate already noted

**Figure 3.1**
**Micronesian Sailing Chart of the Marshall Islands**

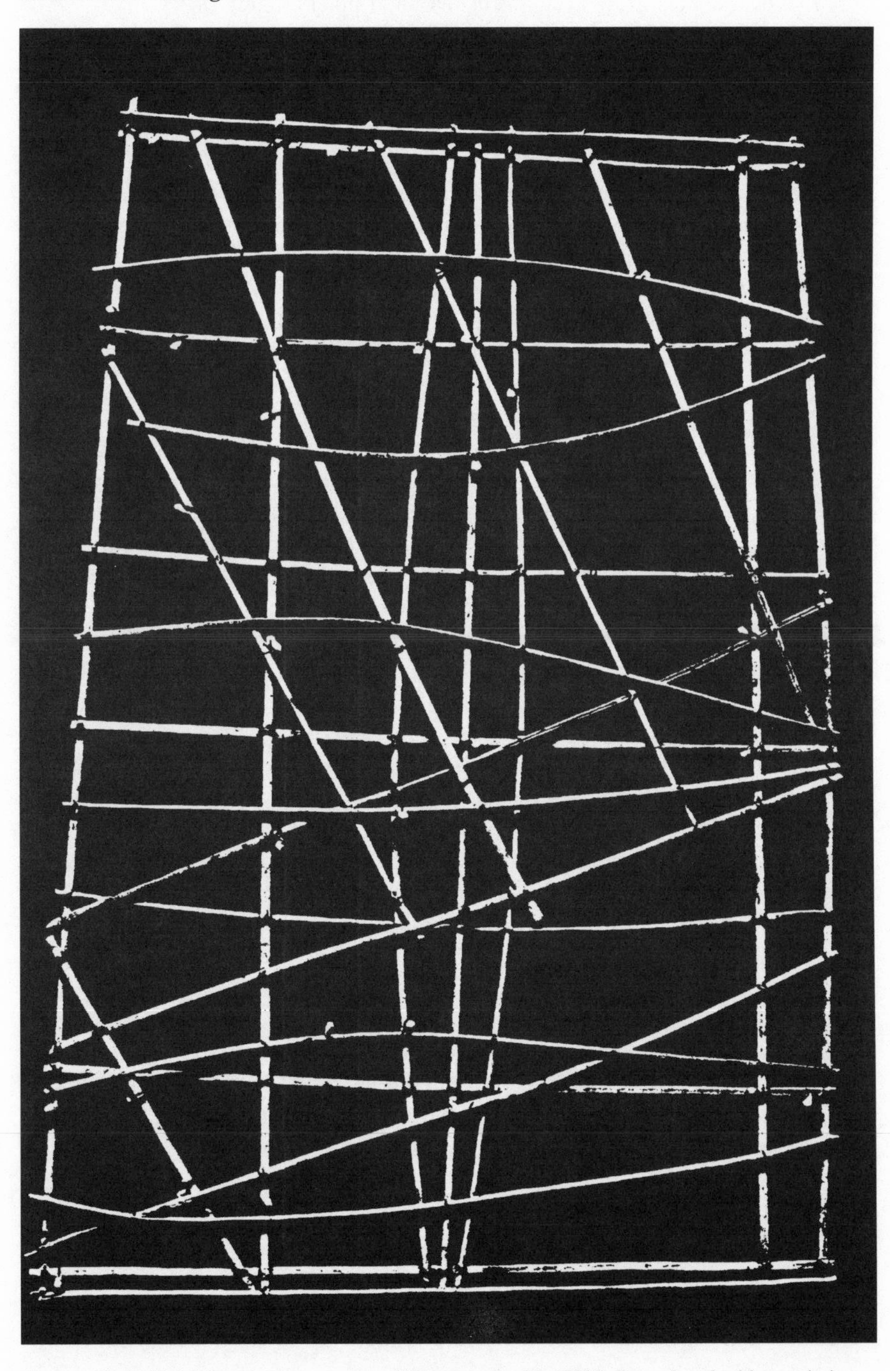

*Source*: The British Museum, Department of Ethnography, No. 2289. Reproduced with permission.

(Nagy et al., 1987). It is important to keep in mind, however, that some children's vocabularies grow at a slower rate, and others, much faster.

It would be a handsome arrangement if the use of a word foretold a reasonably full comprehension of its meaning (or frequently, its many meanings), even among adults. Even there, however, much of any conversation is characterized by attempts to elaborate, and thereby to beg assurance of a complete communication. If words possessed exact meanings, there would of course be much less need to engage in conversations—or for teachers to feel the need to talk as much as they do, a too commonly observed phenomenon of classroom life. In reality, of course, words in and of themselves possess no inherent meaning. Whatever meaning there may be only lies in the mind of the speaker or listener. We use oral and written/printed language to stimulate minds into meaning making. And sometimes—perhaps too often—there is a paucity of meaning despite the number of words flowing forth.

Awareness of the difficulty words pose in expressing or assessing understanding is particularly evident where geographic concepts are concerned because there is a tendency to use technical terms far in advance of the extensive conceptual framework undergirding them. Textbooks are too frequently guilty of using such terms, and teachers buy in to the validity of their use as a consequence, a fact we have known for a considerable length of time. For example, in a delightfully titled piece, "Children's Empty and Erroneous Concepts of the Commonplace," a study report that appeared in 1923, researchers Flora Scott and G. C. Myers pointed out how vague and uncertain elementary school children were of terms and concepts such as "manufacturing," "equator," and even "mile."

Today's situation hardly seems different, despite the advent of television and modems, purported avenues for understanding that have yet to bear their promised fruit, especially where children and young adults are concerned. Concepts are understood as a consequence of relevant experiencing, from which ideas emerge. Words are endowed with meaning by the meaning maker—the reader or thinker, not the teacher, television, or other electronic devices. Consequently, as we utilize geographic terms (e.g., *river mouth, latitude and longitude, region, products, slope*), a certain awareness must prevail. First, constraints function limiting the ability to understand within the general frame of growth and development even when the language necessary to the expression of thought is present; and second, words can only represent, or stand for, thoughts. Piaget alerted us to the first principle, perhaps better than any other, while Tom Robbins put the latter distinction particularly nicely when he wrote, in 1976:

> A book no more contains reality than a clock contains time. A book may measure so-called reality as a clock measures so-called time; a book may create an illusion of time; a book may be real, just as a clock is real (both more real, perhaps, than those ideas to which they allude): but let's not kid ourselves—all a clock contains is wheels and springs and all a book contains is [words in] sentences. (p. 124)

## A WORD ABOUT SEX

The discussion so far has spoken of growth, development, and learning without reference to gender. In the popular culture, we generally assume that girls/women perform less well on spatial matters than do boys/men, no matter their age. In fact, most of the research supports this contention (Ben-Chaim, Lappan, & Houang, 1988). Why this is so is not nearly as clear, but it is generally assumed that the root cause for the observed variances may be assigned to the simple physical fact of sex differences: boys are one way and girls are another. As it turns out, the causes of sex differences, no matter their particular characteristics, are very difficult to determine. As in most research questions, causal relationships that are initially perceived as being rather obvious and as demonstrations of simple truths often turn out to be exceptionally complex and difficult of solution (Harris, 1981; McGee, 1982; Newcombe, 1982; Self & Golledge, 1994).

Of the various ''causes'' most often cited (beyond the naive claim of foreordination), there are two major categories within which researchers have attempted to find causal relationships: (1) assignment of neural/developmental differences between males and females, and (2) evidences of differences in socialization practices. While it is appealing to believe that males are ''hardwired'' to a greater extent than females for spatial awareness, and although startling advances have been made in understanding (and denoting) sex differences in the development of human neural systems, this argument appears to weaken as we learn more about human anatomy. Simplistic notions of laterality and of the meanings the assymetry of the two halves of the brain might have for learning have faded with an increasing appreciation of the complexity and individuality of human behavior. And while we are learning this truth over the broad range of development and learning, it has become increasingly apparent where geographic concepts are concerned (Feingold, 1988; Newcombe, 1982, pp. 228–232).

A much more likely explanation of sex differences lies, I believe, in the differences we observe in socialization. In Western, as in most other cultures, boys and girls are socialized from their earliest years in different ways. We expect girls to engage in sedentary activities; boys are expected to be physically active, searching, and explorative, bringing them into broader contact with the environment. However, it is obviously not a black-and-white situation. Each sex represents a wide range of spatial awareness, and there is much overlap between them. And it is interesting to speculate about the degree to which these ''facts'' might be altered by experience. One sort of experience is obviously formal instruction, and in one study designed to examine this question, it was found that girls learned to visualize spatial relationships equally well (Ben-Chaim et al., 1988). Another study sought to discover what, if any, differences existed in map-reading abilities between children growing up in two countries, Norway and the United States (Douglass, 1972). In both instances, boys outscored girls;

however, Norwegian children outscored their American counterparts, boys over boys and girls over girls. Experiential differences—it is generally conceded that Norwegian children have much greater direct contact with the natural environment—were thought to account for a large measure of the difference between nationalities.

Because experiential factors are much harder to isolate in a research design, it becomes difficult to establish causal relationships. Much is therefore left to inference in evaluating the effect of experience in growth, development, and learning. It is difficult to argue, however, that experience means little or nothing; otherwise we would be hard put to justify much more than rudimentary instruction in our schools.

## OBSERVATIONS FROM RESEARCH

Remembering the caveat regarding Piaget's unfortunate proclivity toward determinism, we remain in debt to him because he was able to think of questions, and ways of getting answers to them, in an entirely novel way. While this is generally true, it is particularly relevant to our quest to discover patterns in children's thought processes basic to understanding spatial interactions. It has generally been assumed that the long-acknowledged failure of students to acquire geographic knowledge and information resulted either from poor teaching, a lack of application in learning, or possibly both. This has been the traditional view, and one can see it imbedded in the National Assessment of Educational Progress, the survey commissioned by the National Geographic Society (U.S. Office of Education, 1990). Piaget, in dramatic contrast, introduced us to the idea of qualitative differences in how children perceive space and spatial relationships. And while the research into children's thought processes in this regard remains incomplete—even spotty and disparate—some general agreements seem to be emerging.

First, the intellectual progression in the ability to comprehend spatial properties moves from perceived space to conceived space, from experiencing space only in the most direct sense to conceptions of space in which the child's thoughts about space, at first quite primitive, gradually become abstract and based on Euclidean (mathematical) conceptions.

Second, conceptions regarding the uses to which space is put, as exemplified by studies of ideas about territoriality, also emerge slowly and developmentally, paralleling the generally perceived patterns found in the child's conception of space. Ideas of country, state, city, town, nationality, foreignness, and so forth become meaningful only in the later period of intellectual development.

Third, it is becoming increasingly apparent that the child's ability to comprehend spatial properties and interactions, while related to intellectual capacity, is more importantly due to the presence of opportunities to access the landscape as well as the ability to manipulate it. In reviewing some of the pertinent research projects, evidence of this principle will, I believe, become self-evident.

## The Basic Piagetian and Related Research Studies

Since Piaget's studies have had such a profound effect on subsequent research efforts to understand how spatial concepts appear to grow and develop, I include here descriptions of the most salient ones, using Piaget's own narrative where appropriate. It is, I believe, an essential background to appreciating how patient we must be in encouraging the development of spatial concepts. It also provides a basis for identifying the kinds of experiences children need, both at home and at school, if they are to grow in this kind of awareness.

### *Conceptualizing the Horizontal and the Vertical*

The following discussion focuses on how Piaget describes research into the steps children appear to follow in gaining control over the concepts of the horizontal and vertical axes (Piaget & Inhelder, [1948] 1956). While mastery of these ideas is likely not a sufficient condition for comprehending the notion of the grid, it doubtless is a key idea as it contributes to the ability to utilize directions. It is equally a component of concepts relating to understanding exact or mathematical locations, and thus of such ideas as latitude and longitude. Further along, it is necessary in comprehending the various concepts relating to distortion and the consequence of rendering phenomena located on a sphere to a flat surface and, perhaps, in appreciating the meanings involved in the concepts of slope and elevation. Piaget and Inhelder describe this experiment as follows (see Figure 3.2 for the typical response for the various substages).

> For the horizontal . . . the children are shown two narrow-necked bottles, one with straight, parallel sides and the other with rounded sides. Each is about one-quarter filled with coloured water and the children are asked to guess the position the water will assume when the bottle is tilted. Some empty jars are placed before the child, the same shape as the models, on which he is asked to show with his finger the level of the water at various degrees of tilt. In addition, the youngest children are asked to indicate the surface of the water by a gesture so that one can be sure whether or not they imagine it as horizontal or tilted. The experiment is then performed directly in front of them and they are asked to draw what they see. Children over 5 (on the average) are given outline drawings of the jars at various angles and asked to draw the position of the water corresponding to each position of the bottle, before having seen the experiment performed. . . . As soon as [the subject] has made this drawing the child compares it with the experiment which now takes place. He is then asked to correct it or produce a new drawing and so passes on to other predictions. (p. 381)

They then turn to the vertical axis:

> Firstly, during the preceding experiment on the jars of water, we floated a small cork on the surface of the water with a match-stick rising vertically from it. The child is asked to draw the position of the "mast" of this "ship" at different inclinations of the jar and then correct his drawing after seeing the experiment. Secondly, we suspended a plumb-

**Figure 3.2**
**Stages in the Development of Horizontal and Vertical Axes**

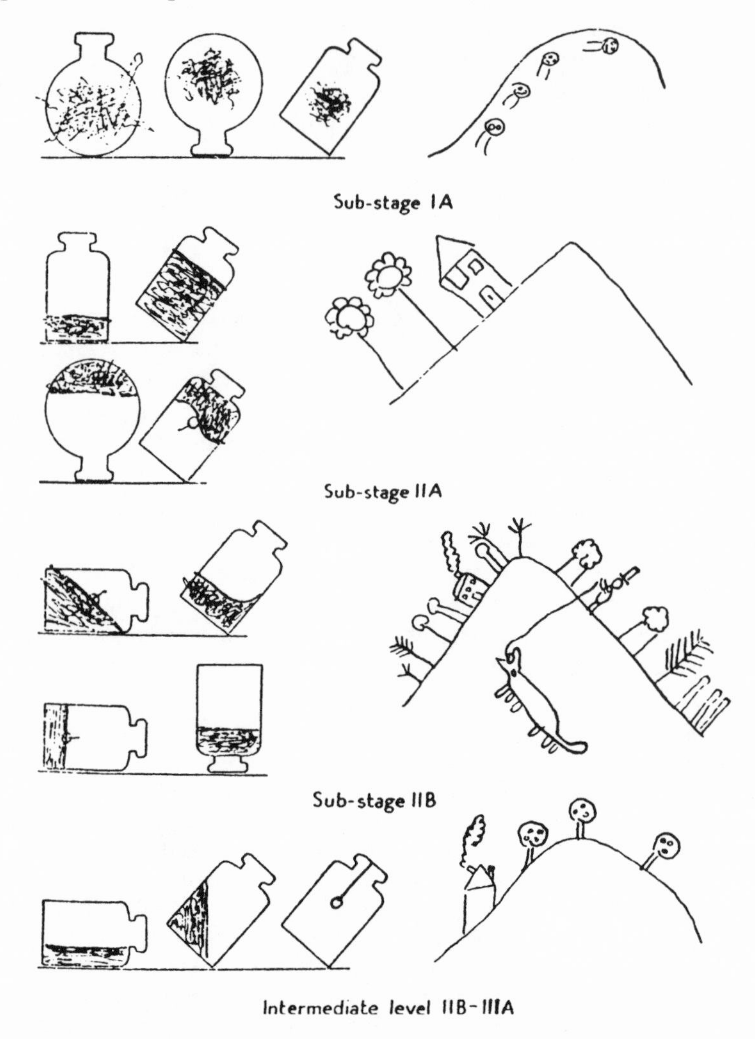


*Source*: Piaget and Inhelder (1956). Reproduced with the permission of Routledge.

line inside the jars (now empty), the plumb-bob being shaped to represent a fish. The child has to predict the line of the string when the jar is tilted at various angles. This done, the experiment of actually tilting the jar is performed and the child is asked for further drawings. Thirdly, the child is shown a mountain of sand, plasticine, etc., and asked to plant the posts "nice and straight" on the summit, on the ground nearby, or on the slopes of the mountain. It is very important to get him to make clear what he means by "straight" and "sloping" in referring to the posts. . . . The child is also asked to draw the mountain, showing the posts either "nice and straight" or sloping. Finally, we sometimes combined the experiment using the plumb-line with that of the mountain, by getting the child to predict the direction of the string when the bob was suspended from hooks projecting from posts planted on the sides of the summit of the mountain. (Piaget & Inhelder, [1948] 1956, p. 382)

Commenting on the findings, Piaget and Inhelder writes:

The first stage lasts until about the age of 4 or a little later. When the children are asked to draw the level of the water in a bottle or the trees on the side of a toy mountain their reaction is extremely interesting, for they are unable to distinguish planes as such. Consequently, they show the liquid neither as a line, nor as a surface, but as a kind of ball (as soon as they get beyond mere scribbling). They think of the fluid in purely topological terms, merely as something inside the jar, and not according to euclidean concepts like straight lines, planes, inclinations, and dimensions. (p. 384)

When the child learns [Substage IIA, 5–6 years] to abstract the surface of the liquid as a plane and locate the trees relative to the mountainside he still fails to grasp the orientation of the water in a tilted vessel or that of the trees to an inclined slope. In the case of the water he thinks of it as moving toward the neck of the bottle, but not by simple displacement. He imagines it as expanding, increasing in volume, and it is because of this increase that it draws nearer the neck as the jar is tipped, while the surface remains parallel to the base. (p. 387).

Midway between the responses of the type described above and the gradual discovery of horizontal and vertical [Stage III, 7–8 to 11–12 years], are a series of reactions which merit careful study. For it is the child's groping attempt [in Stage IIA, 6 to 7–8 years] to solve the problem through persistent trial and error which often provides the key to his subsequent construction of operational systems.

The first step forward occurs when the child indicates on the jar itself the direction in which the water will move when the jar is tilted. At this stage however, his drawings still show the water parallel with the base of the jar just as in Stage IIA. The vertical axis (as shown with the cork floats) also remains perpendicular to the surface of the water regardless of tilt. (pp. 392–393).

Our hypotheses are strikingly confirmed by the results obtained in Stage III [7–8 to 11–12 years]. The children do not discover the vertical and horizontal at one stroke, as might well have been expected had they been unable to give correct answers merely for lack of skill in drawing. At Stage I the child cannot isolate straight lines and planes, and at Stage II he fails to make use of reference systems going outside the pattern he has in mind. At Stage III, however, he begins, though only very gradually, to master more extended reference systems and to construct co-ordinate axes embracing the entire spatial field.

In the course of this process, three main types of reaction may be distinguished. Firstly, one intermediate between substages IIB and IIIA, comprising the discovery of the horizontal when the jar is lying on its side, together with partial discovery of the vertical. Next, Substage IIIA, marked by trial and error construction of vertical and horizontal axes for all positions. Lastly, Substage IIIB, in which this construction is formulated in operational terms and is applied directly to all situations. (Piaget & Inhelder, [1948] 1956, p. 401)

Commenting further on the meaning of Figure 3.3, Piaget and Inhelder write:

> Now at first glance nothing could seem more elementary than a space organized according to such a principle [a three-dimensional system of orthogonal coordinates]. When we view the familiar objects around us, they appear arranged within a grid of parallel straight lines, crossing each other perpendicularly in three dimensions. And if this view of things appears self-evident, it is because physical experience itself seems to force upon us just such a structure, by virtue of all the verticals we perceive as parallel and appearing to cut the [horizontals] at right angles. Indeed, any piece of squared paper, parquet flooring, street crossings, or groups of buildings suggest the same ubiquitous and ineluctable notion of co-ordinate axes. . . . However, the findings . . . show clearly enough that it would be a complete mistake to imagine that human beings have some innate or psychologically precocious knowledge of the spatial surround organized in a two-or three-dimensional reference frame. . . . Far from constituting the starting point of spatial awareness, the frame of reference is in fact the culminating point of the entire development of euclidean space, just as the notions of succession and simultaneity, synchronous and isochronous, defining a homogeneous time, mark the culmination rather than the starting point for the concept of time. (quoted in Flavell, 1963, p. 333)

Goodnow (1977) is one of a number of people interested in children's art (see also Arnheim, 1969; Kellogg, 1967) who has studied this aspect of human development. Her research, too, has shown that many children ignore the "ground line," or horizontal, and the "stand line," or vertical, in their drawings (Goodnow, 1977, p. 7). She observes:

> Some [children] will draw all the members of a family standing on a common ground, but others will draw them in a way that leaves some, to our eyes, floating on air, or even upside-down. Equally common are chimneys drawn at 90 degrees to the roofline rather than to the ground, or people drawn at 90 degrees to a line that stands for a hill. (p. 7)

Only when children can conceptualize the whole can they cope with more than one figure or item at a time, thus bringing a more realistic coordination among relationships when multiple elements are perceived to exist. Hence, we at first see drawings of human beings minus hands or perhaps arms or some other body part, followed by such misconceptions as a right-angle chimney.

### *The Sandbox Experiment*

This study sought to understand how children grow in their ability to represent a known area on a flat surface. An exercise first in modeling, it used smaller objects to stand for the larger reality. Second, it asks the youngster to arrange the objects in a spatial relationship of one to all the others. It involves abilities in comprehending distance and direction of objects in the environment within a restricted frame either in an actual sandbox (an artifact of many classrooms years ago, now rarely seen), or on butcher paper laid on the floor.

Piaget and colleagues describe the experiment as follows (Piaget, Inhelder, & Szeminska, 1960):

[Upon entering the research room] the subject . . . is taken to the window where he is asked to point out various buildings and well-known places . . . merely to ascertain the extent of his local knowledge and sense of direction. Next, he is . . . given a sand-tray with wet sand, carefully levelled off. He is also given a number of wooden houses of various sizes, representing the school and nearby buildings, little pieces of wood representing greens, recreation grounds, public squares and bridges, and a ribbon to represent the [river] Arve. . . . The experimenter takes the biggest house and puts it in the middle of the sand tray, saying: "Now this is the big school (meaning the primary school as against the kindergarten). There are plenty more houses, little ones and big ones. . . . Now I want to know about everything near the school. You put the things in the right places." . . . At the end of the first part of the experiment the subject is asked to draw a plan in the sand or on a piece of paper showing how he would go home from school or, better still, how he would go to a place which they all know. . . . The drawing is a free drawing but the child is asked to show how it fits in which his general plan. Next the child is asked to make a drawing of the sand-model on a large sheet of paper. When he has done this, the experimenter turns the "school building" through 180 degrees and asks: "Now if I turn the school round like this, must we move everything else about as well or can we leave it just as it is?" The child is asked to make the necessary changes himself. With older children the entire experiment can be carried out with pencil and paper.

There are thus three related parts to the enquiry: (1) a plan of the school buildings and the principal features in the immediate vicinity; (2) a reconstruction of the route from school to a well-known landmark; and (3) changes in the location of features when the school building has been turned through 180°. (p. 5)

Findings of this experiment indicate that, in Stages I and II, or up to an average age of 7, "landmarks are uncoordinated and changes of position cannot be described" (p. 7). However, in Substage IIIA (8 to 9;6*), "we find the beginnings of objective grouping . . . and also the beginnings of coordinated landmarks [objects placed in relationship to one another]. . . . However, children at this level are unable to coordinate the system as a whole, either in describing a route or in showing the topographical relations between landmarks" (p. 15). In Substage IIIB (9;6 to 12), "landmarks are fully coordinated, changes of position are represented as a comprehensive group . . . [and children] can construct a topographical schema in line with a two-dimensional coordinate system, though the various intervals are not always strictly proportional to each other" (p. 19).

After reviewing the work of Piaget and Inhelder and related research, Roger Hart constructed a research design intended to refine and extend Piaget's findings (1981). In this instance, he theorized from Piaget's work the existence of three conceptual levels in the development of children's orientation to the landscape: an egocentric system of reference, a fixed system of reference, and an

*The semicolon separates age by year and month.

abstract system of reference. In the egocentric reference system (2–7 years), the child's orientation to space is presumed to be dominated by egocentric thought; the pre-Copernican notion that everything exists only in relation to oneself. In the fixed system of reference (beginning around 7 or 8 years), the child begins to think logically about relationships between objects in the environment. The "child graduates from an entirely egocentric system of reference to a fixed one centered on a small number of uncoordinated routes and landmarks" (p. 200). When the coordinated system of reference begins to take shape (somewhere between 8–9 years and 11–12), adult-like behaviors emerge in which the grid and the cardinal directions serve as anchors for features in the spatial surround.

Like other researchers after Piaget, Hart found a much wider age range in performance. However, he also became much more convinced of the accuracy in Piaget's observation, asserting that "the development of children's spatial activity in their everyday geographic environment, and variations in the freedom of this spatial activity, are the important forces influencing the quality, as well as the extent, of children's ability to represent the spatial relations of places in the large-scale environment" (Hart, 1981, p. 207). Put another way, spatial knowledge is acquired through exploration, or experience with the environment, and not so much through the accumulation of information. He concludes:

> [T]he ability to coordinate larger, more distant areas was [in Hart's research] so markedly influenced by the degree and nature of a child's transactions with the landscape that the influence of the child's general intellectual level does not seem to be the most important factor. . . . [B]ecause the opportunity to explore places freely varies so greatly, the extent of a child's spatial range and such factors as the mode of locomotion through the environment become overwhelmingly important. (p. 207)

In a 1994 book titled, *The Geography of Childhood: Why Children Need Wild Places*, authors Gary Paul Nabhan and Stephen Trimble) tell us:

> The geography and natural history of childhood begins in family, at home, whether that home is in a remote place or in a city. Many naturalists start their journeys on ditchbanks, in empty lots—in any open space just beyond the backyard fence. (p. xiii)

In a series of essays, Nabhan and Trimble suggest how children come to care deeply about the natural world and emphasize the importance of this discovery to every aspect of their lives. They also describe the sadness we must all experience when we realize that it is not just children growing up in the cities and suburbia but even children in the most rural of areas whose contact with reality, such as it is, is a vicarious one derived from endless hours in front of the television set.

### *Decentering—The Mountain Experiment*

One of Piaget's most famous experiments dealt with the concept of decentering. The question in its general form stems from one of Piaget's fundamental concepts, his belief that development, from infancy, is marked by a steady progression away from a strictly egocentric conception of the world, the pre-Copernican view, to one in which the individual is capable of comprehending how others see the world.

The notion of decentering permeates all Piaget's work, but in our particular instance we are concerned with how egocentrism might affect one's perception of a spatial environment. In this particular experiment, the child is asked to take the perspective of a doll viewing a model landscape from different positions (Figure 3.3). Responses are judged by the child's selection of pictures and from pieces of cardboard shaped and colored the same as each of the mountains ''seen'' by the doll. Piaget and Inhelder report the findings (Piaget & Inhelder, [1948] 1956, pp. 210–246) by age/stage, closely paralleling the stages of horizontal/vertical development discussed earlier. Children from 4 to 6 years, 6 months (6;6), or Stage IIA children, were confined to reproducing their own point of view regardless of the position of the doll. Children in Stage IIB (between ages 6;7 and 8 years) were described as in a transitional stage in which attempts were made to distinguish between the different points of view (however, with a tendency to regress to the viewpoint of Stage IIA). Children in Stage IIIA (ages 8 to 9;6) achieved genuine but incomplete ability to shift perspective, while children in Stage IIIB (ages 9;6 to 12 years) obtained complete ''relativity'' of perspective.

Of Piaget's many experiments, this particular one has perhaps been the most subject to criticism. Margaret Donaldson (1978) has taken Piaget to task for creating an experiment in which the language and situation generally are too abstract for younger children to comprehend what is being asked of them. Thus, when the test situation is modified, children who previously were unable to decenter in their thinking became able to do so. Others, particularly the Russian psychologist Lev Vygotskii (1986), have critized the theory along more fundamental lines. Vygotskii, for example, believed that development was characterized by a move away from the social to the egocentric, thus taking much the opposite of Piaget's position.

The argument is important because geographic thinking, including the ability to read maps, requires one to decenter: to imagine or view a landscape and to hypothesize abstract relationships in a spatial environment. If there are limitations on the emergence of this ability, then adjustments within the curriculum might seem in order. But our concern should not simply be, as it was in Donaldson's case, with how much younger children may be able to demonstrate this ability. There is growing concern among those who study the development of cognitive powers that if we follow Piaget's protocols, a significant number of adults, never achieve Formal Operations, the stage during which reasoning escapes the bonds of concrete experience and when it becomes possible to think in hypothetical terms. Perhaps Piaget and others who followed his star have made the same error Donaldson believes has been made where younger children are concerned; it is the vocabulary of the questions that determines who is

**Figure 3.3**
**The Three Mountains**

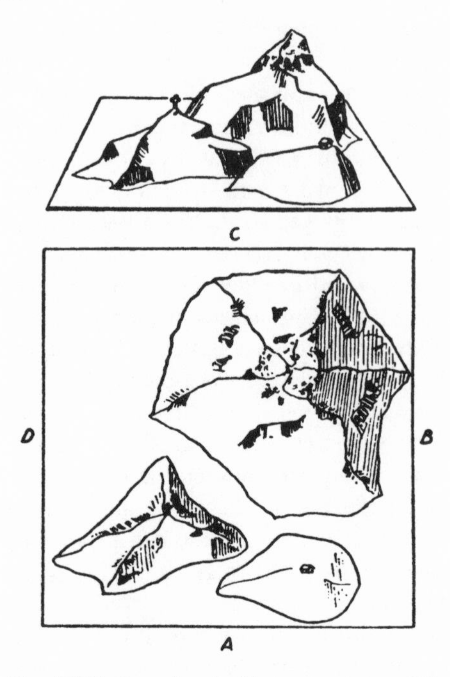


*Source*: Piaget and Inhelder (1956). Reproduced with the permission of Routledge.

''stuck'' somewhere in Stage III and therefore unable to read maps constructed by others.

### *Conceptions of Territoriality*

Piaget's earliest foray into understanding the acquisition of children's spatial concepts is reported in his 1928 book, *Judgment and Reasoning in the Child* (Piaget, [1928] 1968). There he reports an experiment designed to illuminate how Swiss children defined their relationships in a number of different senses: among siblings and other relatives; in defining objects to be to left or the right of; and in conceptions of territoriality involving conceptions of their town or city and their ''decentered'' ideas of political units like one's county, state, and nation. As with the other concepts noted here, Piaget found only a handful of children (about 15%) able to deal with any of these concepts under six years of age. Indeed, the acquisition of ''formal operational thinking'' was required before an accurate conception of territoriality was achieved.

''Are you Swiss?'' [Piaget asks] Very frequently the answer was as follows: ''No, *I am Genevan.*—Then you are Swiss?—No, *I am Genevan.*—Then you are Swiss? No *I am Genevan.*—But your father is Swiss?—No, *he is Genevan.*'' . . . Finally we asked a large number of Swiss schoolboys from all the cantons: ''Are you Genevan (or Vaudois, etc.).'' . . . Up to the age of 9, three-quarters

**Figure 3.4**
**Percentage of Children in Stages I, II, and III by Age (N = 224)**

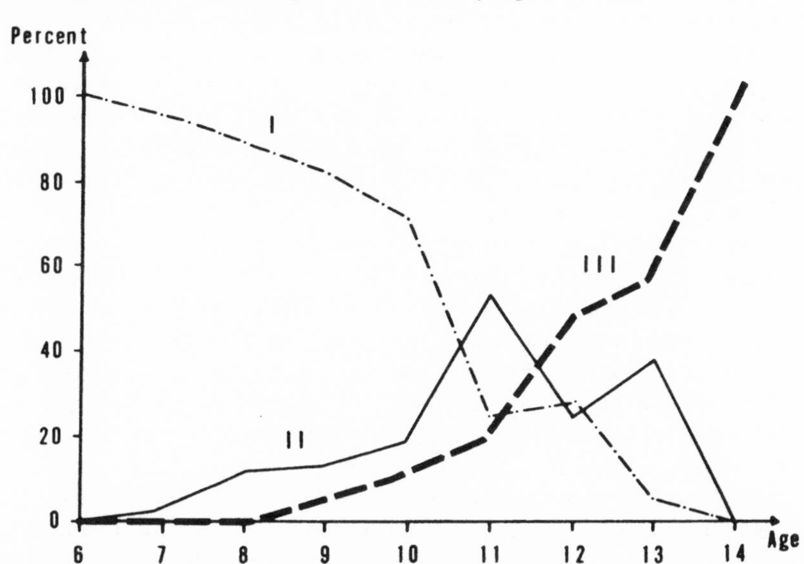


Oliveira's research illustrates the notion that initial achievement of each of the three major developmental stages is likely, for any group of children, to extend over a period of several years.

*Source*: Stoltman (1976). Reproduced with the permission of Western Michigan University Department of Geography.

of the children denied the possibility of being both Swiss and Genevan (or Vaudois, etc.) (p. 119). Piaget found confusions of this sort in almost every combination of questions he managed to ask. A "country" to one youngster is a town to another (at 7;11 years). Another, at 9;5 years, asserts Geneva to be in Switzerland, and Switzerland to be bigger than Geneva, but adds that "you cannot be in both at the same time" (p. 124).

A Brazilian study replicated Piaget's study (Oliveira, 1976) with school children from 6 to 14 years of age. Generally, Piaget's findings were confirmed with one important exception: the Brazilian children generally found at each "stage" covered a much wider age range than Piaget had postulated. "The passage from Stage I to Stage III was uneven, with children moving from one stage to another at different ages" (p. 25) (see Figure 3.4).

Gustav Jahoda, a Scottish psychologist, attempted to replicate Piaget's work on children's conceptions of territoriality (1963). Using children between the ages of six and eleven years, Jahoda identified four geographical levels of understanding. In the first, the children thought of Glasgow, where they lived, as a nearby but otherwise vague entity. "They did not consider Glasgow to include their own location. They described it as being 'beside the playground; up by the park; and, down in Y Avenue' " (cited in Stoltman, 1976, p. 7).

In the second stage, Jahoda (1964) found children to be aware of being part of something called Glasgow and of Glasgow as the place where they lived. However, the idea of Scotland or Britain went undifferentiated. In the next stage, the children were still unable to distinguish beyond the recognition that "Glasgow is in Scotland." "They were not aware of Scotland as an entity within the British Isles, nor of Britain as a national entity. Scotland was identified variously as a town and a country" (Stoltman, p. 8). (Jahoda apparently believed it questionable whether the children's ability to report, "Glasgow is in Scotland," was anything more than pure verbalization). In the fourth stage, children reached adult-like understandings of territorial designations.

Another replication effort (Stoltman, 1976) conducted in the United States compared and contrasted the acquisition of the following territorial concepts: town-city, county, state, and nation. As in the Scottish study, a certain lack of congruence with Piaget's original research manifested itself. Although the children were also found to vary more widely with respect to the age-stage relationship, their responses were similar; as well, the lack of evidence showing the effects of schooling in Piaget's and other studies, must give concern. This is particularly so when one realizes that the Scottish children had undergone considerable formal instruction in geography as part of the standard British curriculum—an interesting contrast with the American one, which rarely provides such formal space for a geographic curriculum. Similar findings are reported by Hart (1981), who attributed much of the confusion among children in the upper elementary grades to a general lack of information. Whether it is entirely a matter of a failure of instruction, as he suggests, needs further analysis. But when such findings are reported from countries where geographic instruction is an important part of the curriculum, we must agree that Piaget may well have identified a reality of human growth.

## MENTAL MAPS AND GEOGRAPHIC THINKING

A novel set of cross-cultural studies provides a particularly interesting introduction to the idea of "mental maps," the images people conjure up about spaces familiar to them that are psychological, rather than Euclidean or mathematical, conceptions of reality (Gould & White, 1986). Research into the nature of the mental maps people hold of places not only gives one ideas about the nature and extent of popular knowledge, it also reveals information about how communication networks of various sorts vary from place to place. "Differences between the attributes of 'here' and 'there,' " the authors write, "have always been of great interest to human geographers, because it is precisely the differences between places that generate movements of goods, people, and information" (p. 1). For example, it is these kinds of mental perceptions that govern the patterns of urban-rural immigrations. They reveal attitudes of peoples of various regions, as in the North-South dichotomy that remains reflected in relations growing out of the Civil War. And, as yet another but more recent phenomenon, with the advent of the electronic age, specific physical locations have lost their various but specific influences. Increasingly, people are free to choose locations to carry out their functions based on other criteria: perhaps an environment free of pollution, one in which schools are notably superior, or even a locale where recreational facilities are especially attractive.

Educationists may be particularly interested in the uses of mental maps to better understand "worldviews," even if they are of a rather restricted portion of the entire world of school-age children. Information revealing the extent of children's knowledge of their spatial surround has also been elicited from individual children, as is shown in Figures 3.5, 3.6, and 3.7 (from Gould & White, 1986). In this instance, we see the maps of three black youngsters living near the Mission Hill area of Boston, an almost exclusively white area at the time the research was carried out. The students were first asked to draw a map of their "area"; they were then interviewed (and tape-recorded for later analysis). In these examples, "Dave" includes the residential Mission Hills Project but has a large blank area on his map. In fact, he views the project as a forbidden area, one he is afraid to enter. As a consequence, it looms large, and undifferentiated, whereas the areas familiar to him include considerable detail. We also see how Parker Street appears as a boundary/barrier between the world he knows best and the no-man's land just beyond. "Ernest" also sees such a boundary/barrier. The scale he has employed makes it appear as an even greater psychological barrier, and in fact the researcher discovered that neither of the two had ever physically ventured into the white enclave lying on the other side of Parker Street. "Ralph" has a different perspective. A student at the prestigous Boston Latin School, he not only physically travels farther than the other two students, but the detail in his map reveals interests not present in the other two. For example, he has broadened his scale so as to include five educational institutions in the area, indicating his interest in education as an escape from the life of the black ghetto in which he lives.

The third example (Figures 3.8–3.10) reports the differing perceptions of adults from three different socioeconomic areas of Los Angeles (from Gould & White, 1986). In the first, and most detailed (Figure 3.8), we see Los Angeles perceived through the eyes of upper-middle-class whites in Westwood. In the second (Figure 3.9), the view is from the largely black south-central Los Angeles area, while the third (Figure 3.10) reports the area as seen from the mostly Hispanic, eastern portion of the city. Interpreting the meanings derived from these three examples, Gould and White comment:

> Upper class, white respondents from Westwood had a very rich and detailed knowledge of the sprawling city and the wide and interesting areas around it, while black residents in Avalon near Watts had a much more restricted view. For the latter, only the main streets leading to the city centre were prominent, and other districts were vaguely "out-there-somewhere," with no interstitial information to connect them with the area of detailed knowledge. Most distressing of all was the viewpoint of a small Spanish-speaking minority in the neighborhood of Boyle Heights. This collective map includes only the immediate area, the City Hall and, pathetically, the bus depot—the major entrance and exit to their tiny urban world. (p. 17)

**Figure 3.5**
**Dave's Map**

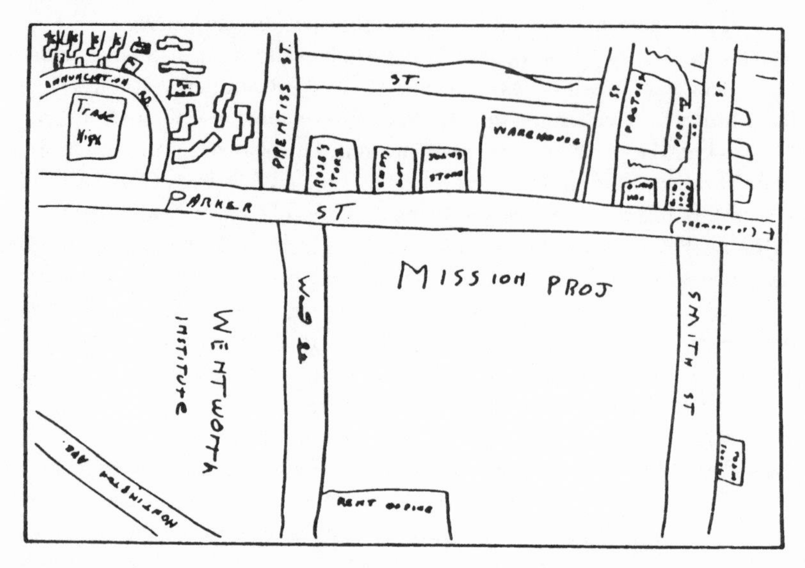


*Source*: Gould and White (1986). Reproduced with the permission of Routledge.

**Figure 3.6**
**Ernest's Map**

*Source*: Gould and White (1986). Reproduced with the permission of Routledge.

**Figure 3.7**
**Ralph's Map**

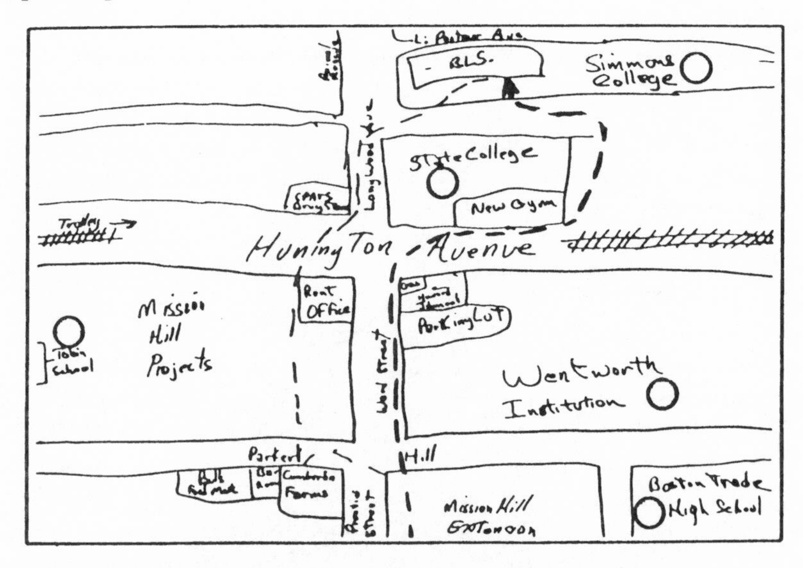


*Source*: Gould and White (1986). Reproduced with the permission of Routledge.

Another study of the mental maps of prospective teachers adds a dimension to the problem of teaching for geographic literacy (Chiodo, 1993). A group of students preparing to become elementary and secondary teachers and enrolled in a social studies methods class were asked to draw, as rapidly as possible, a map of the world. They were to be as accurate as they could manage in a 20-minute period. Their product maps were evaluated according to the apparent ability of the students to draw and label the seven continents, to draw them in the correct hemisphere, and in their approximate relative size and location to one another.

Not surprisingly, those preparing to teach in secondary schools outperformed their elementary school counterparts and, as is generally the case, men outperformed women. However, when the same mental map–drawing activity was administered to seventh graders, it became virtually impossible to tell the difference between their maps and those of the teachers-to-be, in a seeming replication of a 1968 study (Drumheller) in which, amazingly, little difference was found between maps produced by sixth graders and by college sophomores. Further, an analysis of the backgrounds of students in the Chiodo study showed that above-average scores were obtained by those students who had taken more than one college geography class, who read and enjoyed reading geographic material (including the *National Geographic* magazine), and who had traveled more than their lower-scoring counterparts.

Once again, we are reminded of the importance of direct experiencing of the environment to the development of spatial understandings.

**Figure 3.8**
**Los Angeles through the Eyes of Upper-Middle-Class White Residents of Westwood**

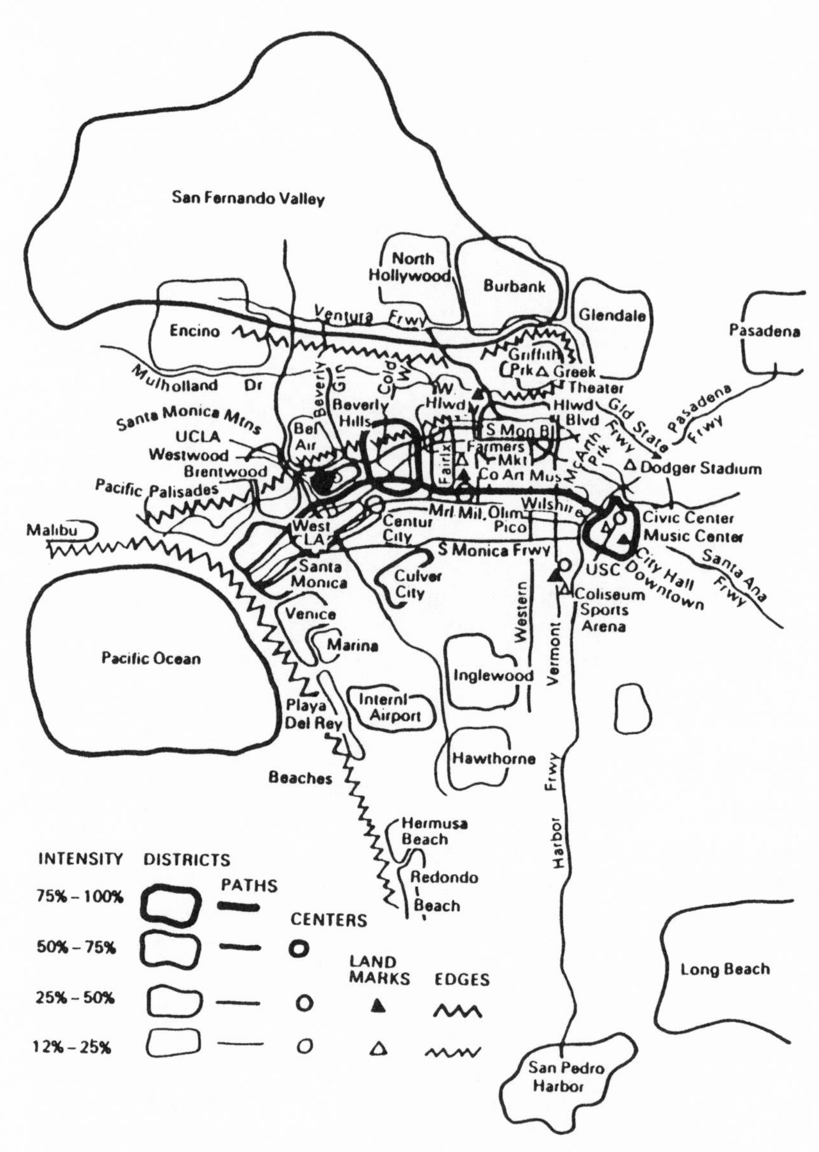


*Source*: Gould and White (1986). Reproduced with the permission of Routledge.

Figure 3.9
**Los Angeles through the Eyes of Black Residents of Avalon**

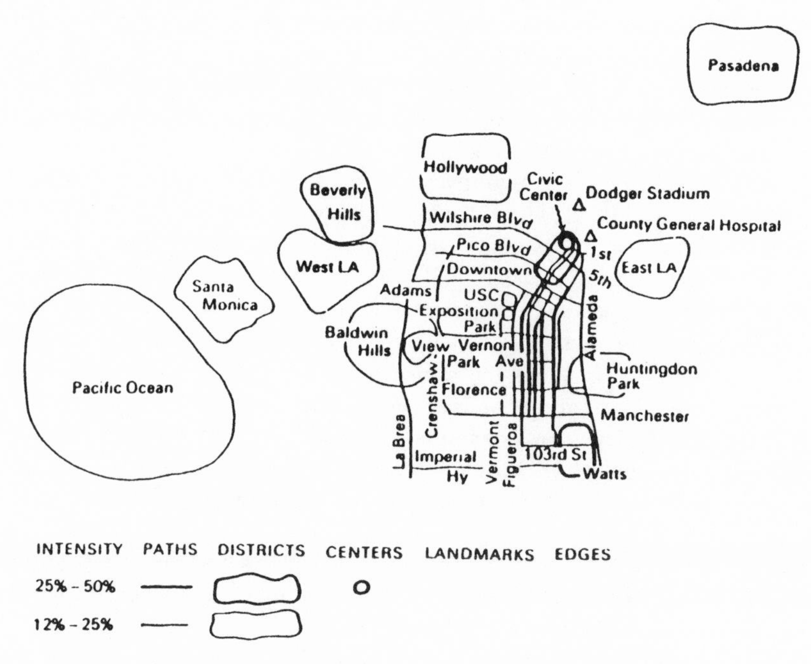


*Source*: Gould and White (1986). Reproduced with the permission of Routledge.

Figure 3.10
**Los Angeles through the Eyes of Spanish-Speaking Residents of Boyle Heights**

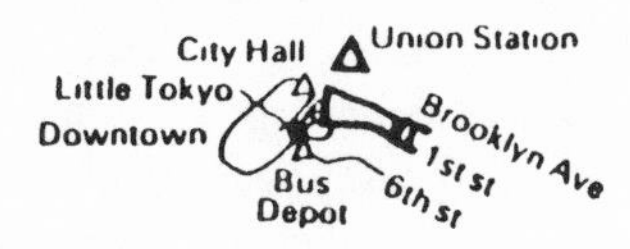


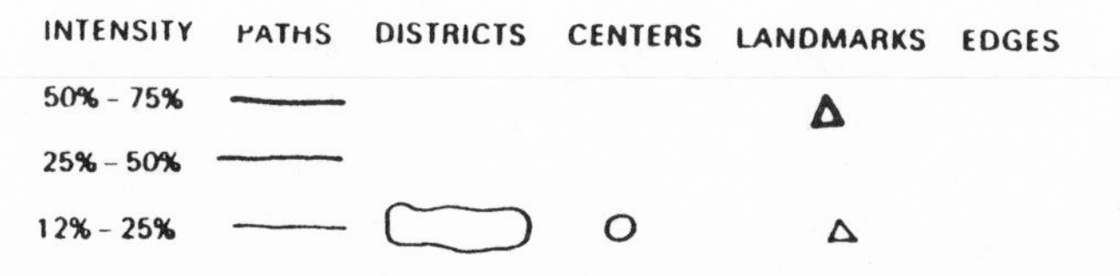


*Source*: Gould and White (1986). Reproduced with the permission of Routledge.

## IMPLICATIONS FOR GROWTH AND DEVELOPMENT, TEACHING, AND LEARNING

When children enter the formal educational system, they generally have a considerable amount of spatial experiencing behind them. Most know their immediate physical environment well, and they have begun constructing what might be called cognitive maps of that environment. They are able to conceptualize, albeit largely by intuition, a spatial environment, which enables them to move about with relative freedom. Within that space, they know how to get from A to B, and they are able to reverse their knowledge of such relationships, thus finding A from B. They are also learning how to go from A to B and from B to C. Soon thereafter, they comprehend how to take detours, how to go from A to B by alternate routes, and similar feats. Early on, getting to and from school, to nearby stores, and to a friend's house may become seemingly natural accomplishments, yet these are all actually ''learned'' abilities derived not from formal instruction, or intuition, but from directly experiencing the environment.

Although there exists meager research evidence informing us of the patterns children follow as they learn to control, manipulate, and understand their spatial environment, it seems clear that elementary-age children need to develop their own personal frame of reference before they can conceptualize spatial representations. That is, they require extensive experience in relating themselves to the immediate spatial environment before they can think productively about spatial relationships within an abstract frame of reference, as is required when attempting to read a map devised by someone else. Put another way, it appears that *the first stages in learning to think geographically derive from highly personal and extensive experiencing with the immediate environment, a complex process, that is ongoing through most, if not all, of the so-called elementary school years*—until the age of 11 or 12 in most instances. How do such conclusions match our teaching strategies in the classroom and at home?

If firsthand experiencing of the environment is as crucial as the research suggests, then we have a considerable problem on our hands, as a majority of our elementary school–age children live in urban or suburban communities. Direct contact with the environment, the importance of which has been emphasized repeatedly from Piaget's time onward, is often and at best spotty. Today's child is transported hither and yon, coming to the ground for different kinds of contacts and activities in first one place and then another. Since these descents from the automobile usually occur miles apart, the chance for the child to develop a cohesive sense of a spatial environment seems remote. If we are to take at all seriously the need for such a conception of one's environment in building a foundation on which larger worldviews are built, then we can sense the dimensions of the problems that urban/suburban living presents.

I have emphasized the child's transactions with the environment, a factor I believe to be more important than rigid designations of intellectual growth, although some people will disagree. One more example of the role such experience plays in development can be found in a fascinating study conducted in Devon, England (Lee, 1963, cited in Hart, 1981), in which it was found that children who had previously walked to school after being bused experienced social and emotional adjustment problems. Commenting on the effects of busing in his own study, Hart believed it confirmed, "most dramatically . . . that children younger than 8 years of age cannot (with rare exceptions) incorporate the school bus journey into their [own maps or] topographical representations" (p. 216).

What might the school's role be in building a better foundation for geographic literacy? One immediately thinks of the possibilities the neighborhood school might exploit in helping elementary school youngsters achieve a sense of spatial wholeness within the home-school environment. However, this means conceiving of the classroom as something much larger than the space to which today's children are typically confined. A curriculum that does not cultivate a familiarity with the spatial characteristics of the school buildings themselves and the site on which they are situated is, of course, failing its clientele. This represents only a beginning, however. The new curriculum must also incorporate ways of bringing firsthand experiences with the environment beyond those confines, to the home-school-community beyond the boundary of the schoolyard. Alas, most curriculums conceive of the classroom as the only environment within which children are to learn, so we rarely see pupils engaging in any kind of systematic learning outside its four walls.

School policies militate specifically against venturing outside the classroom door. Growing concern that the school avoid exposing itself to issues of liability encourages teachers to keep their charges within the classroom. Perhaps a more insidious pressure derives from an apparently exploding demand for formal assessments of academic progress. Faced with requirements, from administrators and parents, for "accountability," teachers must resist a strong tendency to abandon teaching opportunities that show little promise of translating directly into improved test scores. Hence we see in such programs an overwhelming emphasis on basic arithmetical competencies and on print reading. It is little wonder teachers report so little time being spent on instruction in the other curriculum areas (see Chapter 2). Yet another such influence (a more subtle one) is emerging as public concern over the general effectiveness of the schools is growing. This comes in the form of curriculum designs promulgated at the state level to counteract the increasing pessimism over the ability of the schools to respond to what is perceived to be a generally ineffective institution. Especially where the social science curriculum is concerned, I call the reader's attention to a public school curriculum termed the California Social Studies Framework. One might normally disregard, and justly so, a particular state's efforts to design a curriculum to meet its own needs. In California's instance, however, we see a state that has been influential nationally beyond all proportion to the rest of the country on virtually every social, political, economic—or educational—matter one might mention. In the matter of the California History-Social Science Framework (California State Board of Education, 1997), one can expect a great deal of anticipation regarding the influence it will exert in other parts of the nation. And in this respect, persons interested in the geographic aspects of the social science curriculum will quickly realize the potentially negative effect of the heavy emphasis on history contained within. Written as a reaction to critics who decry the lack of specific information in students' minds and intended to countervail long-held opinions regarding the intellectual vapidness of ''the social studies,'' the framework reflects public policy pressures more than it does a sound conception of the interrelationships among the social sciences (indeed, as described in Chapter 2, neither history nor geography are exclusively ''social sciences'' since both concern themselves as well with matters in the natural sciences, the arts, and so forth).

From the discussion so far, the reader will appreciate the importance I place on the early years in establishing the groundwork for the emergence of formal geographic thinking. Although prior to this time by which we know children have been exposed to a multiplicity of geographic ideas, thoughts, references, facts, and information about world events and places, we cannot be at all sure there has been the kind of retention that we presume adults—also, perhaps, equally hopefully, but almost as inaccurately—garner from the potpourri that is everyday television fare. (One eight-year-old, for example, in describing how a nationally televised football game came to an end, asserted that the players, whom the child confused confusing with the announcers, ''all got dressed and went back to New York.'' Indeed, it would appear that in this instance, as throughout the language acquisition process, children incorporate many ''geographic'' terms into their vocabularies. They use these words, as they do the rest, in ways that suggest the possibility of understanding but that, on further examination, reveal notions that are either inaccurate or naive. One would be well advised to proceed on the assumption there is less there than meets the eye (or ear).

Nonetheless, later childhood reveals an emerging Euclidean sense of order, and it is from this point onward that the school can effectively introduce ideas basic to the appreciation of a physical world on which human beings continue to exert their willfulness and, often, their artlessness. Because the map is the basic tool of the geographer and of anyone else engaged in geographic thinking, facility in reading maps is essential to the geographic curriculum. In the next chapter I deal with the process of reading as a basic form of behavior and how our knowledge of it applies in learning to read the language of maps. I have attempted to show in this chapter how, it is believed, children grow toward understanding the Euclidean concepts that underlie mature map reading. These developmental steps I believe to be fundamental to achieving the skills and abilities that mature map read entails. We omit or pass over them in the earlier school years, to the detriment of our students. Without an awareness of space in these earlier forms, students can hardly be expected to make sense of the kinds of concepts the reading of true maps entails.

But what if, as children reach adolescence and the secondary school years, they have not had the kinds of spatial experiences described here? We do not know the answer with any exactitude, of course. However, one must wonder how much reliance can be put on haphazard experience to build the kind of background that more formal learning would seem to require. The alternative relies on a curriculum for the younger years that takes the student into the real world. Unless we provide the opportunity for direct experiencing, we may very well be handicapping secondary school teachers in their attempts to help students build a systematic comprehension of the world.

## REFERENCES

Arnheim, Rudolph. 1969. *Visual thinking*. Berkeley: University of California Press.

Ben-Chaim, David, Lappan, Glenda, & Houang, Richard T. Spring 1988. The effect of instruction on visualization skills of middle school boys and girls. *American Education Research Journal*, 25.1, 55–71.

Bruner, Jerome A., Goodnow, Jacqueline J., & Austin, George A. 1956. *A study of thinking*. New York: Wiley.

California State Board of Education. 1977. *History-social science framework for California public schools, kindergarten through grade twelve*. Sacramento: California State Department of Education.

Chiodo, John J. May/June 1993. Mental maps: Preservice teachers' awareness of the world. *Journal of Geography*, 92.3, 110–117.

Donaldson, Margaret. 1978. *Children's minds*. New York: Norton.

Douglass, Malcolm P. 1972. The development of map reading abilities: A cross-cultural perspective. In F. J. Monks, W. W. Hartup, & J. de Wit, eds., *Determinants of behavioral development*. New York: Academic Press, pp. 539–545.

Drumheller, S. J. 1968. Conjure up a map: A crucial but much neglected skill. *Journal of Geography*, 2, 140–146.

Feingold, A. 1988. Cognitive gender differences are disappearing. *American Psychologist*, 43, 95–103.

Flavell, John H. 1963. *The developmental psychology of Jean Piaget*. New York: D. Van Nostrand Co.

Gesell, Arnold. 1930. *The guidance of mental growth in infant and child*. New York: Macmillan.

Gesell, Arnold. 1940. *The first five years of life: A guide to the study of the preschool child from the Yale Clinic of Child Development*. New York: Harper & Brothers.

Gesell, Arnold, & Ilg, Frances L. 1946. *The child from five to ten*. New York: Harper & Brothers.

Gesell, Arnold, Ilg, Frances L., & Ames, Louise Bates. 1956. *Youth: The years from ten to sixteen*. New York: Harper.

Goodlad, John I. 1984. *A place called school: Prospects for the future*. New York: McGraw-Hill.

Goodnow, Jacqueline. 1977. *Children drawing*. Cambridge, MA: Harvard University Press.

Gould, P. 1965. *On mental maps*. Michigan Inter-University community of Mathematical Geographers No. 9. Ann Arbor: University of Michigan Press.

Gould, P. 1983. Getting involved in information and ignorance. *Journal of Geography*, 82.4, 158–162.

Gould, Peter, & White, Rodney. 1986. *Mental maps*. 2nd ed. Boston: Allen & Unwin.

Harris, Lauren J. 1981. Sex related variations in spatial skill. In Lynn S. Liben, Arthur H. Patterson, & Nora Newcomb, eds., *Spatial representation and behavior across the life span: Theory and application*. New York: Academic Press, ch. 4.

Hart, Roger A. 1981. Children's spatial representation of the landscape. In Lynn S. Liben, Arthur H. Patterson, & Nora Newcombe, eds., *Spatial representation and behavior across the life span: Theory and application*. New York: Academic Press.

Heatwole, Charles A. March/April 1993. Changes in mental maps. *Journal of Geography*, 92.2, 50–55.

Jahoda, G. 1963. The development of children's ideas about country and nationality. *British Journal of Educational Psychology*, 33, 47–60.

Jahoda, G. 1964. Children's concept of nationality: A critical study of Piaget's stages. *Child Development*, 35, 1081–1095.

Kellogg, Rhoda. 1967. *The psychology of children's art*. New York: Random House.

Lee, T. R. 1963. On the relation between the school journey and social and emotional adjustment in rural infant children. *British Journal of Educational Psychology*, 27, 100.

Liben, Lynn S., Patterson, Arthur H., & Newcombe, Nora., eds. 1981. *Spatial representation and behavior across the life span: Theory and application*. New York: Academic Press.

McGee, Mark G. 1982. Spatial abilities: The influence of genetic factors. In Michael Potegal, ed., *Spatial abilities: Development and physiological foundations*. New York: Academic Press, pp. 199–222.

Nabhan, Gary Paul, & Trimble, Stephen. 1994. *The geography of childhood: Why children need wild places*. Boston: Beacon Press.

Nagy, William E., Anderson, Richard C., & Herman, Patricia A. Summer 1987. Learning word meanings from context during normal reading. *American Educational Research Journal*, 24.2, 237–270.

Newcombe, Nora. 1982. Sex-related differences in spatial ability: Problems and gaps in current approaches. In Michael Potegal, ed., *Spatial abilities: Development and physiological foundations*. New York: Academic Press, pp. 223–252.

Oliveira, L. 1976. The concept of territorial decentration in Brazilian school children. In J. P. Stoltman, ed., *International research in geographic education: Spatial stages development in children and teacher classroom style in geography*. Kalamazoo: Western Michigan University Department of Geography.

Piaget, Jean. [1928] 1968. *Judgment and reasoning in the child*. Totowa, NJ: Littlefield, Adams & Co.

Piaget, Jean. 1950. *The psychology of intelligence*. New York: Routledge & Kegan Paul.

Piaget, Jean. 1954. *The construction of reality in the child*. New York: Basic Books.

Piaget, Jean, & Inhelder, Barbel. [1948] 1956. *The child's conception of space*. Reprint. London: Routledge & Kegan Paul.

Piaget, Jean, Inhelder, Barbel, & Szeminska, A. 1960. *The child's conception of geometry*. New York: Basic Books.

Piaget, Jean & Weil, A. 1951. The development in children of the idea of the homeland and of relations with other countries. *International Social Science Bulletin*, 3, 561–578.

Pick, Herbert L., Jr., & Lockman, Jeffrey J. 1981. From frames of reference to spatial representations. In Lynn S. Liben, Arthur H. Patterson, & Nora Newcombe, eds., *Spatial representation and behavior across the life span: Theory and application*. New York: Academic Press, pp. 39–62.

Rand, D., Towler, J., & Felhusen, J. 1976. Geographic knowledge as measured by Piaget's spatial stages. In Joseph P. Stoltman, ed., *International research in geographical education*. Kalamazoo: Western Michigan University, pp. 57–78.

Robbins, Tom. 1976. *Even cowgirls get the blues*. New York: Houghton Mifflin.

Ross, Dorothy. 1972. *G. Stanley Hall: The psychologist as prophet*. Chicago: University of Chicago Press.

Scott, Flora, & Myers, G. C. November 1923. Children's empty and erroneous concepts of the commonplace. *Journal of Educational Research*, 8, 327–334.

Self, Carole M., & Golledge, Reginald G. September/October 1994. Sex-related differences in spatial ability: What every geography educator should know. *Journal of Geography*, 93.5, 234–243.

Stoltman, Joseph P. 1976. Children's conception of territory: United States. In Joseph P. Stoltman, ed., *Spatial stages development in children and teacher classroom style in geography: International research in geographic education*. Kalamazoo: Western Michigan University Department of Geography, pp. 39–56.

Towler, J., & Price D. 1976. The development of nationality and spatial relationship concepts in children: Canada. In Joseph P. Stoltman, ed., *International research in geographic education*. Kalamazoo: Western Michigan University Department of Geography.

U.S. Office of Education. February 1990. *The geography learning of high-school seniors*. Washington, DC: U.S. Office of Education.

Vygotskii, Lev. 1986. *Thought and language*. Translated and edited by A. Kozulin. Cambridge, MA: MIT Press.

Wadsworth, Barry. 1978. *Piaget for the classroom teacher*. New York: Longman.

# 4

# What It Means to "Read" a Map

Today we are a society where our kids know the floor plan of Nordstrom better than the map of the world.

U.S. Senator Dianne Feinstein, in a speech to supporters in Century City, California, October 5, 1992

A map . . . is a testimony of a man's faith in other men; it is a symbol of confidence and trust. It is not like a printed page that bears mere words, ambiguous and artful, and whose most believing reader—even whose author, perhaps—must allow in his mind a recess for doubt.

A map says to you, "Read me carefully, follow me closely, doubt me not." It says, "I am the earth in the palm of your hand. Without me, you are alone and lost."

And indeed you are. Were all the maps in this world destroyed and vanished under the direction of some malevolent hand, each man would be blind again, each city be made a stranger to the next, each landmark become a meaningless signpost pointing to nothing.

Yet, looking at it, feeling it, running a finger along its lines, it is a cold thing, a map, humorless and dull, born of calipers and a draughtsman's board. That coastline there, that ragged scrawl of scarlet ink, shows neither sand nor sea nor rock; it speaks of no mariner, blundering full sail in wakeless seas, to bequeath, on sheepskin or a slab of wood, a priceless scribble to posterity. This brown blot that marks a mountain has, for the casual eye, no other significance, though twenty men, or ten, or only one, may have squandered life to climb it. Here is a valley, there a swamp, and there a desert; and here is a river that some curious and courageous soul, like a pencil in the hand of God, first traced with bleeding feet.

Here is your map. Unfold it, follow it, then throw it away, if you will. It is only paper. It is only paper and ink, but if you think a little, if you pause

> a moment, you will see that these two things have seldom joined to make a document so modest and yet so full with histories of hope or sagas of conquest.
>
> Beryl Markham, *West with the Night*, pp. 245–246

## WHAT IS "READING"?

When we use language in everyday living, we employ the word *reading* in its generic sense, meaning that there is available in the environment the widest possible range of "stimuli" that may be read. For example, we commonly speak of reading music, diagrams, signs and symbols of various kinds, formulas, charts, pictures, faces, maps, colors, situations, comments of others ("I read you"), weather, tracks—and, of course, words. Some of the many things we say we read—at least out of school—we read rather well; others we read not so well, or perhaps not at all. Failure to read something in this broader sense can occur because of an oversight, but more often the reader is bereft of the conceptual frame, the experiential background, that is necessary if sense is to be made of the new stimuli available for "reading." Without that reservoir of experience, the reader will lack the power to propel his or her thinking processes to new levels of understanding. In this generic, or broader, sense, then, we use the term *reading* to identify the process by which one creates meanings—namely, by which one expands one's experience base—for those things in the environment for which some sort of awareness can be generated.

In its generic sense, the behavior we call reading can engage any stimulus form. Some of the things available for our reading call for what one might describe as direct, or primary reading. These are the sorts of things that occur naturally in the environment—clouds or other atmospheric phenomena, the nature of soils, characteristics of the landscape, marks left on the environment by human activity, and so forth. A second form encompasses what we might term indirect, or secondary reading. These are the signs and symbols that humans have created to stand for things that are observable in the primary environment or that stand for ideas that are abstractions of other ideas that ultimately relate to our perceptual world: words, formulas or equations of various kinds, and other signifiers of reality, such as those occurring on maps and globes. Some of the things available to us to read as we seek to make sense of maps, for example, appear as fairly obvious abstractions of what might be found in its natural context as, for example, various kinds of human structures (buildings, bridges, railroads). Others are much more abstract: words themselves, lines connecting points of equal value (*isobars* on a weather map, showing points of equal barometric pressure, and *isolines*, indicating points of equal elevation, for example), and others showing a flow of some sort (trade, population, wind patterns).

## HOW WE LEARN TO READ

How one learns to read these sorts of stimuli is not a matter on which there is universal agreement. In "real life," meaning for the moment life out of school, while we may speak of "reading" as reasoning or thinking, in most school situations the term takes on a much narrower definition. In this context, we think of it only in terms of responding to printed words. In addition, we reverse what we accept as obvious in everyday living (but without much analysis)—that form follows function. In learning that is not circumscribed by what we think transpires in a formal educational situation (in *school*, for example), most would agree that one develops skill-like competence as a consequence of experience. For example, in learning to swim or ride a bicycle, practice is the harbinger of competence, perhaps with a modest assist here or there from someone who has developed the essential skills at an earlier time and who has the capability of analyzing the behavior and a knack for helping someone else at that teachable moment, when an improvement in a particular aspect of form (or skill) translates itself into improved function.

But these are not intellectual tasks, you say. When it comes to intellectual competence, can we say that the situation changes qualitatively as well as quantitatively? Children acquire knowledge of their spatial environment largely if not perhaps exclusively through experience. But without question, the most amazing example of learning in which form follows function is found in the language acquisition process itself. Here, in a few short years, the infant, who is born with the capacity to speak any of the many hundreds of languages found around the world, somehow eliminates the ability to make those sounds that do not comport with those found in his or her own language community and then proceeds to gain mastery over the grammar and syntax of that first or native language in a mere six or seven years, meanwhile acquiring a vocabulary of several thousand words, the size of which will continue to increase more or less indefinitely.

Until relatively recently, we thought that children acquired this ability to speak (and to listen, although with considerably less acumen) largely by imitating adult models. This assumption has been carried over into the school curriculum, where we have traditionally provided the intellectual models, or forms, assuming that function would follow. In the case of that area of the curriculum most subject to worry over whether students will achieve mastery, the ability to read print, for example, we have consistently believed that fluent or mature reading behavior can only result when students are first taught the skills that we associate with such behavior. Although the problem of teaching reading (in its narrow sense) consumes the elementary school curriculum and budget and takes up an inappropriate amount at the secondary level, where we devote much time to remedial and corrective measures in the hope of making up for the failures of earlier learning, the principle of teaching form first is found throughout the curriculum. It is true where arithmetical (or mathematical) functions are concerned just as it is within the geographic curriculum, where devotion to definitions and locational questions dominates the thinking about what is important to know.

The traditional view of learning as mastery of form is not surprising, given the still widely held conception that children are in many, if not most, respects simply miniature adults. We see evidence of this, even though we may reject it theoretically, in the provision of "dumbed-down" instructional materials, such as books that present the classics for children (rewritten with a presumably simpler vocabulary) and, indeed in all the instructional materials designed to instruct children in print reading (although the same cannot be said to be true of social studies texts, which are frequently loaded with terms far beyong the learner's comprehension).

A major problem for anyone emphasizing the form-following-function paradigm is that those aspects we think of as "function" (i.e., ideas) lack the specificity of "form" (or skills). Standardized testing, which came into being in the 1920s (and now is in an apparently robust, old age), has always been more successful in measuring skills than ideas. A growing dependence on these kinds of measures in evaluating our schools has lengthened the life of instruction that emphasizes skill learning as the core of the educational enterprise.

## IMPLICATIONS FOR MAP READING

Several consequences flow from our present understanding of the nature of (print) reading behavior where map reading is concerned. Map reading entails a particularly complex lexicon of signs and symbols, all appearing within a multidimensional frame of reference. Contrasting map reading with print reading reveals it to be a behavior requiring much more complex thought processes than does the simple reading of a consistent, and relatively limited, group of symbols. As a consequence, it is only realistic to expect development toward fluent map reading to be a much more extended process than we might normally expect where print reading is concerned.

Paralleling developmental patterns evident in the emergence of spatial concepts (described in Chapter 3) with the growth of verbal abilities leads us to several ideas. We can view the preschool years, for example, as a time during which the child explores and comes to know through firsthand experience, an environment with spatial characteristics. How limited or enriched this particular environment may be is obviously dependent on a number of things, including the opportunities that caregivers provide.

When children come to school, the richness of these experiences is something the teacher ought to assess. Just because children roam rather widely over a physical space, we cannot assume that they possess more than a rudimentary acquaintance with its significant characteristics within a context of spatiality. And it is at this point that we should find intentional ways of expanding on this spatial frame of reference. But there are barriers to overcome here, if only because the classroom is by definition a confined space. Young children, and students of all ages, need opportunities to conceptualize spatial relationships on a larger but still personal scale. For younger children, the school grounds and buildings and the immediate community itself provide many opportunities to see things within a spatial frame. For older students, the immediate environment is a larger area, but the objective remains the same. Saying this, I am only too aware of the unfortunate strictures being placed on movement outside the classroom. Legal, financial, and safety concerns increasingly inhibit many good curricular practices, not least of which involves leaving school grounds.

Taking another cue from research in language development and learning, it is also appropriate to emphasize the singular importance of beginning to learn map reading through the writing of one's own map. In the teaching of the language arts based on experiential learning (as contrasted with skills-first instruction), teachers rely on preexisting writing abilities as a basis for introducing the child to the writing of others. Beginning with the dictation of stories to go with the child's paintings, it is but a short step to understanding that it is possible to read, not only one's own stories but also stories that others have created. The concepts a child must learn in acquiring print-reading abilities are, I believe, paralleled in the acquisition of map reading:

> What I can think about I can talk about.
>
> What I can talk about can be expressed in writing (in drawing or some other form of writing).
>
> Anything I can write I can read.
>
> I can read what I write and what other people write for me to read. (Allen, 1961, p. 60)

As in the affinity between writing and (print) reading, I believe the writing or creation of maps should garner the lion's share of the emphasis in the early years, at least through the ages of 8 or 9. Block play provides one of the earliest forms of representing spatial environments. Even children younger than two will engage in block play in a systematic way, and older children become adept at very complex representations of spatial environments. Reading maps "written" by others, while appropriate from time to time, clearly will depend on learning not only the form maps take, but their function as well. This is to say, while the primary objective of the early stages in the development of map-reading ability is to establish the idea of the function of maps, as spatial representations with meanings extending far beyond the mere matters of *location* or descriptions of *place*, maps are not simply works of art. The *form* that maps take ultimately is of importance, and the skills implied in meeting this criterion in mapmaking are of no small consequence.

It is important, therefore, in the writing or creation of maps, for students to begin sensing the notion of the grid as the tool through which accuracy of representation is accomplished. The concept of the grid, ultimately as represented by degrees of latitude and longitude, begins with an understanding that is fundamentally much more complex than it first appears. Awareness of the concept of the grid commences with evidences of its existence in the immediate environment. Classrooms and school buildings, schoolyards, house lots, streets, and roads (in many but not all instances) are laid out as grids. Discovering how to plot a right angle, the importance of measurement, and evidences of slope can be combined to introduce students to concepts associated with the grid and show how this tool helps the mapmaker, and consequently the map reader, to more accurately represent and interpret the spatial environment.

Unfortunately, as in the teaching of print reading, much of what we do within the school's subject matter curriculum elevates the role of form (skill) over function (meaning). Fortunately, however, recent developments in the print reading curriculum promise a reprieve from the imbalance that has affected school literacy efforts in years past. As students have opportunities to learn to enhance their reading and writing skills through practice rather than direct instruction, more time will become available for the consideration of geographic ideas. We can only hope that this development, so long overdue, will take on increased speed.

## CREATING MEANINGS FROM MAPS: A PARTICULARLY COMPLEX FORM OF READING IN ITS GENERIC SENSE

The process by which one becomes a fluent reader of maps created by others presents much the same set of issues as learning any of the other forms of reading. There is, however, a particular caveat: map reading requires the reader to synthesize an unusually large number of specialized concepts, some of which present themselves through symbol systems appearing on the map itself (which itself may be one of a number of uniquely different forms), while others can only be inferred. It thus can be viewed as a particularly complex form of reading in its generic sense.

Before considering the subtleties that come to play in any act involving the interpretation of maps, however, it is essential first to remember that the purpose of any map is to tell a series of stories about human and natural interactions within a specifically confined space: the area the cartographer has chosen to depict. Whether the reading is meaningful and the degree to which it is accurate result quite independently from the efforts of the person who has prepared the map or told the story. As in any other kind of reading, the richness or fullness of any effort to make sense of, or create meanings for, the symbols and relationships depicted is dependent on what the reader brings to the reading itself.

The generation of meanings clearly cannot take place in an intellectual vacuum. Rather, it is contingent on the presence of what educationists call "an appropriate experiential background," meaning the ideas or concepts one has generated in prior living. While hardly anyone would disagree that learning should commence "where the learner is" (a restatement of "appropriate experiential background"), it is often difficult to know just where each student might be for the particular task at hand. This is especially so where map reading is concerned, since here the reader must integrate many different symbol systems, which must be synthesized within a grid, a process that (especially if the map is of a large area) causes actual relationships to vary from place to place within the area depicted.

Generally speaking, we can divide the types of concepts required in mature map reading into two broad categories: (1) those symbol systems that depict human and natural features in various degrees of abstraction, and (2) aspects hidden from direct view: those dealing with distortion, direction, and scale.

## Symbol Systems Basic to Map Reading

Looking at virtually any map, we see that it contains a wide variety of signs and symbols, from print itself through various other kinds of markings and, perhaps, the employment of an entire range and shading of colors. Unfortunately, there is no standard language in map making; each mapmaker elects to use those signs and symbols which it is believed will make the most readable map, given its intended purpose. As part of each map, the cartographer appends a legend, a kind of glossary to which one must refer if any degree of meaningfulness is to result. Consulting the *map legend* does not, of course, guarantee any particular level of understanding. It may in fact be more confusing than enlightening if the reader is bereft of a sufficient understanding of, for example, what a city might look like or (an even more capricious thought), how sizes of cities might be compared and otherwise thought about.

Since the maps most frequently found in classrooms may employ dozens or even scores of different signs and symbols, their use is subject to a great deal of confusion. Who has not heard of the youngster who expressed amazement at finding that a state that was always represented on the physical maps of the United States as pink is not that color at all, nor can the boundary between it and the adjacent state be seen, as previously thought. (A similar misconception, harder to root out, has lived in the minds of many children who thought that the lands of the British Commonwealth, formerly known as the British Empire, were in fact also pink.)

It is perhaps gratuitous to call attention to the fact that the map legend serves only as a handy guide for the person who already has a sufficient background to appreciate the meanings lying behind the symbols, and that without such an experiential base, it is really quite useless. However, when one considers the wide variety of signs and symbols utilized by mapmakers, the problem of map reading becomes quite daunting for the person who is helping students build the experiential background necessary to bring meaning to the map in front of them.

We will start here with the more obvious aspects of map reading: understanding the nature of those signs and symbols that are readily observed when one peruses a map with the thought of making some sense out of the experience. Doubtless, the most obvious of these are the words and numbers (feet, meters, fathoms) themselves, those designating the names of cities, rivers, mountain peaks and passes, notable elevations of these and other phenomena, route numbers and mileage distances (if a road map), and similar items.

A second, and much more diverse, set of symbols includes those that also designate the locations of natural and human phenomena of various kinds. Some of these take on the general appearance of the thing itself (such as the location of an airport, church, bridge, or dam), whereas others are totally arbitrary abstractions. Still others designate real and abstract phenomena: the course of a river, a railroad line, roads and highways, or boundary lines of various levels of political entities.

A third set, involving symbols denoting how certain phenomena are distributed over an area, includes a very wide range of possibilities. For instance, land-use patterns may be depicted to indicate regions within which specific kinds of phenomena are dominant. The uses to which land is put as depicted in a land-use map, (farming, forestry, mining, manufacturing, etc.) comprises one such application. Alternately, one might wish to indicate differences in natural vegetation that occur over a specific region. Weather maps present the reader with a series of lines that connect points of equal barometric pressure. Other maps may be drawn with the specific intent of showing variations in average rainfall, areas of urban versus rural populations, the presence of communicable diseases, and so forth. Maps showing the relative extent/strength of human or natural movements of various kinds—immigration/migration patterns, trade activity, the relative forces of wind and sea movements—are based on the principle of showing dominant or relative patterns of various sorts. Maps of this nature are all created along the principles of the isobar and isoline, which connect points of equal or predetermined value or intensity of a particular phenomenon, whether natural or human. Because mapmakers frequently use colors to distinguish between intensity levels, confusions involving the colors of states or countries are not uncommon.

A fourth element entering the equation involves the depiction of relief and elevation, in other words, the degree of slope occurring naturally in the landscape. Modern methods of measurement (many of which employ aerial photography, frequently from satellites,) make it possible to construct maps depicting relief and elevation with extraordinary accuracy. The most common result is a contour map, on which points of equal elevation (or value) are connected. More recently, land/satellite ("landsat") maps, obtained through sophisticated, satellite-borne, remote-sensing devices, make possible maps that go far beyond merely recording the most obvious physical features in the landscape. Most of the materials available to schools and much of what we see in the popular press, however, eschew these kinds of detail in favor of generalizations, which often hide, and even distort, reality. For some reason, adults preparing materials for students of all ages think that these forms of "simplification" provide something that is more easily understandable, despite the fact that great quantities of specific information—representing knowledge essential to reaching the generalization in the first place—are deliberately omitted. For example, in one such instance, on maps that purport to show landforms directly on the map surface it becomes necessary to exaggerate the actual relief features many times over. This is because the actual physical features of the earth's landscape, unless the mapped area is severely limited, are simply too slight in the overall to be noticeable. Showing elevations directly on the map, even those depicting areas that encompass just a few square miles and certainly those covering much larger areas, always results in distortions of relative landform to area relationships; this is particularly so when large areas are depicted.

Even more difficulty in meaningful reading arises when relief and elevation are suggested by combining colors (to represent elevation) with semi-artistic renderings, which can only suggest differences in relief or slope from place to place. This is probably the most popular form of "instructional map" used today, not just in our classrooms but in everyday life—for instance, it is the universal method employed on the route maps one finds in every airplane seat pocket. A rather mysterious aspect of these "air-brushed" maps lies in the decision to represent the shadow effect necessary to highlight both relief and elevation, as the sun would do in real life, by showing the source of such light as situated in the northwest—a total improbability, at least in the northern hemisphere. Thus we see a major difficulty with maps that combine a form of artistic license with the more precise world of the cartographer: this is another situation in which the reader must deduce specific information from what has become a generalization. This is no small task even for the mature mind; it is daunting for the relatively immature learner.

## Hidden from View: Concepts of Distortion, Direction, and Scale

It is self-evident that the skin of even a carefully peeled orange cannot be made to lie correctly on a flat surface. The same conclusion should obviously be at hand when we think of the problem any mapmaker has in attempting to render a portion of the earth's surface on a surface that is flat. However, the ramifications that follow in mapmaking are not only endless, they are extraordinarily subtle.

For example, immediately following World War II, when Senator Joseph McCarthy began propagandizing about the danger of communism, both from within the U.S. own government and from without, he routinely made his public appearances before a map that included the Soviet Union. The map he chose for this purpose was the famous Mercator Projection, beloved by navigators because it allows them to plot their courses as straight lines. But to do so, the surface areas of the equator are made increasingly extended, with the greatest distortion appearing at the poles. In employing the Mercator, McCarthy was able to point to a map on which the Soviet Union appeared 250 percent larger than it actually is in relation to the rest of the world. The Mercator Projection, useful as it has been since it was first invented in the sixteenth century, serves especially well as an example of the worst type of map where the problems of scale and distortion are concerned. Scale, or the ratio of distance on a particular map to its actuality on the earth's surface, can be seen here to vary progressively toward its extreme the farther removed the reader's eye moves from the equator. The result is a degree of distortion of the size and shape of the land mass that is very difficult to discern through direct examination. Instead, the reader must infer the degree of distortion, a not inconsiderable intellectual feat given that the only direct means of comparison is found in a chart inserted somewhere on the map that depicts scale relative to the degrees of latitude north or south of the equator.

Cartographers have waged a constant battle attempting to minimize the problems brought about by our orange peel analogy. We speak of the various "projections" that have been invented to manage them, referring to the way features of the model globe might appear were they "projected" from its center through the surface of the globe and onto a flat piece of paper. The Mercator Map (or Projection) obviously represents one of the earliest attempts at mapmaking based on this principle (see Figure 4.1). Variations on this theme appeared in subsequent years. Still, the accuracy with which mapmakers were able to construct their maps was limited by their access to direct knowledge about the location of things in the real world. Throughout the early years of cartographic development such information came only from direct observation, involving an often-tedious process of compiling data over an extended period of time.

The march of science has almost totally changed the face of cartographic work, particularly over the recent past. Satellites are the new explorers, relaying all kinds of information to computers that construct maps, many of which represent (and re-present) new categories of information. New knowledge—such as the fact that the earth is not a perfect sphere and, indeed, is constantly changing shape, though in extremely subtle ways—has, among many other recent discoveries, altered conceptions about earth relationships. And still the information-gathering process continues. The earth seems to offer an endless source for discovery of all sorts of information, not all of it exotic—as the crew of the ship *QE2* discovered after supposedly accurate charts led it to run aground on a previously unknown shoal in admittedly treacherous waters off Martha's Vineyard, Massachusetts.

Thus, where we previously relied on direct observation in recording spatial phenomena, today's maps are drawn from inferential data obtained through advanced techniques, much of which originates from satellites. Since the early

**Figure 4.1**
**Central Projection of a Globe on a Cylinder and a Subsequently Modified Map Structure, the Mercator, Made to Same Scale along the Equator**

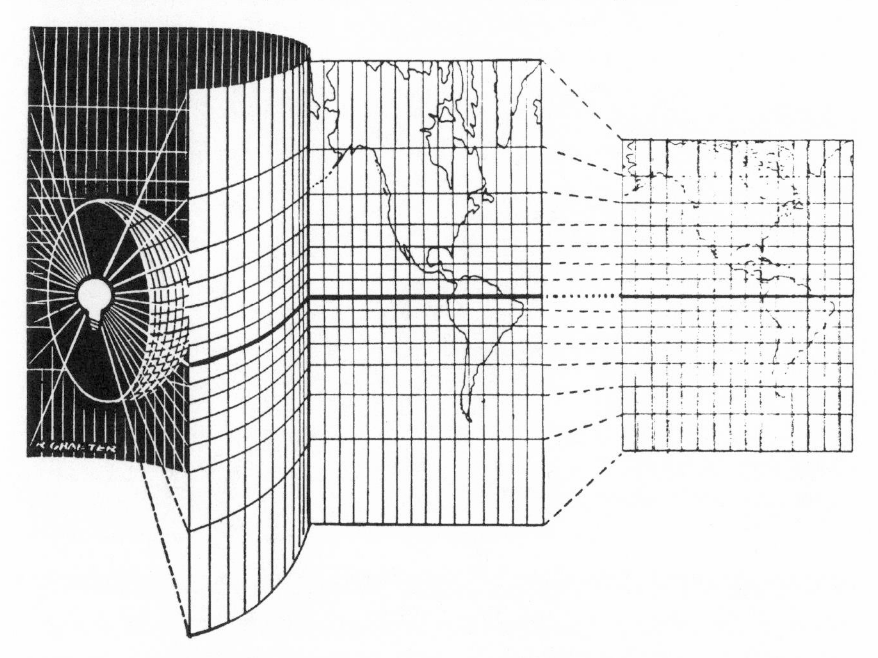

The genesis of the Mercator Projection (and others depicting portions of the earth's surface based on this form of projection) is illustrated here. The projection is "errorless" at the equator, and distortion increases as one moves toward the poles. In actuality, however, the light would never reflect an image in the extreme ranges; hence, Mercator and others who have worked with this projection compensated for the inherent distortion.

*Source*: Greenhood (1964). Reproduced with the permission of the University of Chicago Press.

1960s, the instrumentation used to explore the universe has been applied as well to study the composition of our own planet. *Remote sensing*, as the technique is called, employs various frequencies of the spectrum, which can be read to inform us of the quality and extent of a virtually infinite range of human and natural phenomena. Landsat maps have enriched our knowledge in endless ways, from evidence, for example, of human activities of all kinds (including the movement of military forces) to the presence and richness of natural resources.

The availability of an almost infinite number of ways of representing the earth on a flat surface brings to our attention once again to the problem of how we orient ourselves in space. A sense of directionality is clearly important in map interpretation. It is also an ability that appears to be part of the human psyche. We remain unsure to what degree, or even whether, instruction has any pro-

**Figure 4.2**
**Detail of an Australian Map of the World, 1990s**

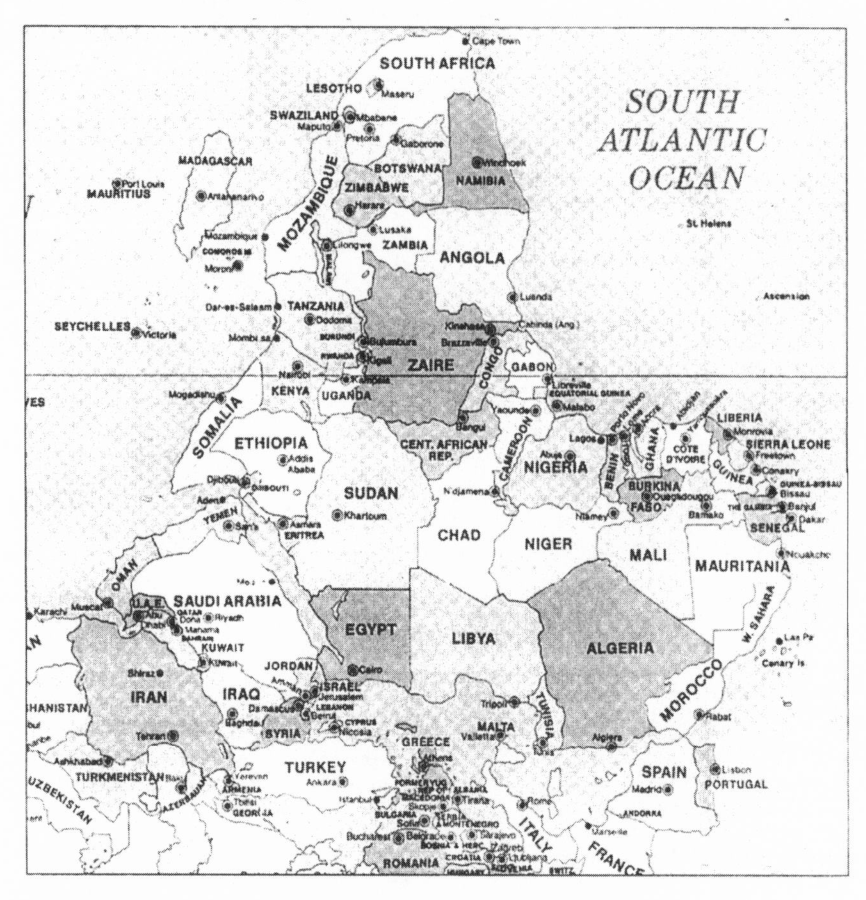


*Source*: Universal Press Pty. Ltd. Used with permission.

nounced effect on one's personal ability to quickly or accurately sense directions "in the wild." What we can be sure of is the misleading nature of the orientation of virtually every map to be found in American classrooms, on which north is invariably at the top. In reality, the so-called polar directions (north, south) have absolutely nothing to do with the top or bottom of anything. They exist quite independently of those concepts, yet the near-universal association of the two makes it next to impossible for us to recognize familiar landforms—such as the outline of countries, continents, states—unless they are presented within the top-to-bottom ("north/south") context. Consequently, most people find it not only disconcerting but totally disorienting to be faced with a map in which this protocol has not been observed. Figure 4.2, for example, shows a portion of a map used in Australian schools. For inhabitants of the northern hemisphere, Africa looks totally different from the way it is usually experienced in American classrooms. Add the lettering identifying the various countries (at least at the time of writing), and one experiences yet a further disorientation. Examples such as this suggest the need to present images of the world (i.e., maps), in various orientations continuously through the school experience.

Cartographers have traditionally sought a number of different solutions to the problem of distortion. Since it is true that the larger is the surface area to be shown on the map, the greater the problem of distortion, one of the more obvious solutions is to include the smallest such area that still relates the story the map is intended to tell. Various solutions to the problem of showing large portions of the earth's land mass, such as a large country like the United States or a hemisphere, have been invented. And although modern maps are now the result of mathematical constructs, they still are based on these different principles. It is in the depiction of the largest areas, including the entire world, that we see the most difficult problems in distortion. One solution is found in *equal area maps*, in which the cartographer spreads the distortion equally over the entire surface of the map, as contrasted with the Mercator Projection, where distortion increases as one approaches the poles. However, the equalization of distortion may also be limited to the land masses, the consequence of which is to make the oceans appear considerably larger than they actually are (in addition to the already extant distortion of the coastlines).

There is obviously no single satisfactory solution to the problem of distortion. Because the purpose of the map in large measure determines the size of the area to be encompassed, the question is not whether there will be increased distortion relative to surface measurements. Rather, it becomes one of how to minimize its effect in reading as accurately as possible the story the map is intended to relate.

One of the most telling examples of the degree to which a map distorts reality is to compare "great circle" routes on the globe with the seeming shortest route between two points on a flat map. Great circle routes can be found for any two points by stretching a string from one such point to the other across the surface of a model globe. Then, by noting the intervening points of latitude and longitude on the globe and translating them to the flat surface of the map, a connecting line will display the "true," linear course or direction, which now appears more like a sine (periodically oscillating) wave than a straight line (see Figure 4.3).

## MAPS IN THE CLASSROOM

The problem of distortion arises in the elementary or secondary school classroom in several ways. I am speaking now of the "pull-down" maps, large "actual relief" maps, and others used primarily in group discussion situations. The first thing to strike one is their common orientation, with north at the top

**Figure 4.3**
**Great Circle Arching a Rhumb on the Mercator**

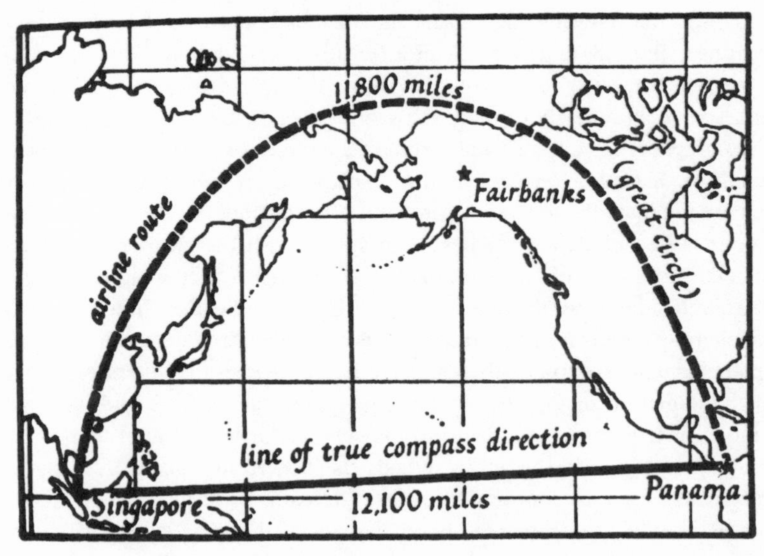


*Source*: Greenhood (1964). Reproduced with the permission of the University of Chicago Press.

of the map. A second circumstance full of potential difficulties for their readers is the mixing of various kinds of uses that these maps are intended to serve. It is not at all unusual to find a political map next to a land-use map, which in turn is next to a map showing physical features. In each case, heavy reliance is placed on colors to represent the intended emphasis. Consequently, if the teacher expects to utilize the map for something more than merely a locational tool, it is very likely that students will fail to "cue in" to the very different meanings that coloration necessarily implies. Without this sensitivity, the consequence may be a misreading in which erroneous concepts are generated or a trivialization of the vast amount of different kinds of data available for reading. Students therefore need to have these differences recalled often enough to bring about automatic recognition of the importance of the map legend in helping them achieve the mind-set appropriate to each particular map.

A multitudinous number of maps devoted to specialized topics appear in textbooks and the many other resources we would hope students might consult in the course of their studies. The following special topic maps are, for example, listed in *The Map Catalog* (Makower, 1992).

**Travel Maps**

Bicycle Route Maps

Mass Transit Maps

Railroad Maps
Recreation Maps
Tourism Maps and Guides

**Maps of Specific Areas**
County Maps
Foreign Country Maps
Native American Land Maps
State and Provincial Maps
United States Maps
Urban Maps and City Plans
World Maps

**Boundary Maps**
Boundary maps (international, foreign countries, state, and local)
Congressional District Maps
Land-ownership Maps
Political Maps

**Scientific Maps**
Agriculture Maps
Geologic Maps
Natural Resource Maps
Topographic Maps
Wildlife Maps

**History Through Maps**
Antique Maps and Reproductions
History Maps
Military Maps

**Utility and Service Maps**
Business Maps
Census Maps
Emergency Information Maps
Energy and Utilities Maps

**Water Maps**
Nautical Charts and Maps

Ocean Maps

River, Lake, and Waterway Maps

**Sky Maps**

Aeronautical Charts

Astronomical Charts and Maps

Weather Maps

**Images as Maps**

Aerial Photographs

Space Imagery

**Atlases and Globes**

Political

Physical

Political-Physical

Every classroom should, of course, contain a model globe against which comparisons can be made with the flat map. High school social studies classrooms should also have available a device that demonstrates the concepts lying behind the basic map projections. Additionally, classrooms should be equipped with indoor/outdoor thermometers, wind vanes, barometers, and rain gauges. Recent developments in electronics provide an exciting opportunity to make available highly sophisticated and accurate instruments in these latter categories at a relatively inexpensive cost.

## A NOTE ON LATERALITY AND KNOWLEDGE OF DIRECTIONS

One's personal orientation in space is generally held to strongly affect, either positively or negatively, the ability to read maps in a meaningful way. Although we have little research evidence that this is the case, we do know that what we popularly call ''sense of direction'' does vary considerably from one person to another. We also know that this personal sense of directionality can either facilitate or impede the ability to function within a particular physical environment.

It has been popular to argue that directional capacities are inborn—either an apparently hopeless burden or a particular talent is sealed into the genetic makeup and is thus finitely determined. The ability to instantly know left from right or east from west has also been extrapolated as playing an important role in learning generally, and particularly in learning to read print. Thus, we can see that deterministic ideas are not yet completely dead.

Such deterministic ideas draw on a long history and an ever-present human fascination with how the brain functions. In the mid-1800s, phrenologists studied the configuration of the head, particularly the protuberances, thereupon inferring a subject's behavioral characteristics. Although we are no longer bound by such primitive assumptions, misunderstanding exists between what we know and what we think is known about how the brain seems to work. The discovery of the fact the human brain consists of two assymetrical hemispheres has led to a great deal of speculation about what may be the unique role of each hemisphere in learning and thinking. For example, it is said that "the left hemisphere is specialized for language, mathematics, detailed analysis, logical thought, temporal and sequential analysis, and serial processing of sensory information." Meanwhile, the right hemisphere is home to "emotional expression, intuition, the recognition of faces and the emotions expressed in faces, artistic achievement, attention, recognition of musical passages and other musical aptitudes, visual-spatial analysis, and parallel processing of sensory information" (Efron, 1990, p. 1). Put more simply, the left hemisphere is where the verbal functions reside, while many nonverbal functions are found in the right hemisphere. Although today very few in the medical specialties accept these dichotomies as true harbingers of behavior, educational practitioners and others with little background in neurology have leapt to embrace the concept of brain asymmetry as a causal factor in explaining human thinking processes, including spatial awareness.

Conjectures about what might be going on if the roles presumably assigned to each hemisphere fail to become clearly established have raised the possibility in unschooled minds that we might be on the verge of explaining why directionality, for example, is not as clearly established in some individuals as others. Similarly, they believe we can explain why some children have difficulty in learning to read print, which does indeed involve directional matters, at least to a certain extent (although we should remind ourselves that the printed form of languages does not always require left-to-right or front-to-back reading, but may involve such variations as right-to-left and back-to-front reading, among other possibilities).

While it is possible to assign certain functions to specific areas of one or the other of the hemispheres from a physiological point of view, these are almost entirely capacities that do not involve thinking as such. In fact, as those in the forefront of neurological research will only too quickly admit, "we do not at present understand the cognitive function of *any* brain area" (Efron, 1990, p. 27). In other words, *thinking* remains largely a mystery, and so attempts to explain thought processes in terms of specific hemispheres, let alone particular portions of the "left" or "right" side of the brain remains beyond divining. Still, a host of articles and books in recent years have touted the importance of "hemispheric specialization" as a significant element in planning for teaching, while asserting it as a raison d'être for what has gone wrong in the educational establishment generally (Hunter, 1976; Edwards, 1979; Harris, 1985).

Perhaps the most charitable thing that might be said about any relationship that may exist between the organization and structure of the brain and one's orientation in space, including the ability to read maps intelligently, is that just about everything awaits proof of its existence. Meanwhile, the best antidote is the opportunity for spatial experiencing. But what is needed is not just experiencing for its own sake, but experiencing that occurs in an environment of planned awareness. The teacher/parent must help the child learn from experience by heightening perceptions gained through the shared interactions with the environment that school and a constructive home life should provide. Table 4.1 lists useful teaching tools and resources.

**Table 4.1**
**Teaching Tools and Resources**

*Discovering Your Life-Place: A First Bioregional Workbook*, by Peter Berg. A guide to mapping the defining characteristics of your home bioregion. San Francisco: Planet Drum Foundation (P.O. Box 31251, San Francisco, CA 94131), 1996.

*How to Draw Maps and Charts*, by Pam Besant and Alaistair Smith. One of the How-to-Draw series, with very useful, well-illustrated guidelines for mapping projects. Tulsa, OK: Usborne/EDC Publishing, 1993.

*Make It Work: Maps—The Hands-On Approach to Geography*, by Steve Watts. One of the best new books on using hands-on geography techniques with children. Elegantly illustrated, with very accessible graphics. Toronto, Canada: World Book, 1996.

*Mapping: ESS Teacher's Guide*, by Beth Barth. The guide illustrates diverse ways to get children involved in making two- and three-dimensional, topographical, classroom, and neighborhood maps. Nashue, NH: Delta Education (P.O. Box M, Nashua, NH 03061–6012), 1985.

*Maps and Mapping: Geography Facts and Experiments*, by Barbara Taylor. Good activities on mapmaking and map reading. Extensive graphics and beautiful illustrations. New York: Kingfisher, 1993.

*Mapstart, 1, 2, & 3*. by Simon Catling. A "series of graded map-skill and atlas books and related materials that gradually develop the basic skills children need to be able to understand and actively use plans and maps of all kinds." Essex, England: Collins-Longman. 1985.

*Selborne Celebration: Annual Journal of the Selborne Project*. A collection of works from schools involved in this program, studying "one square kilometer" around their schools. Jamestown, NY: Roger Tory Peterson Institute (311 Curtis Street, Jamestown, NY 14701), 1996.

*Valley Quest: Quest Maps and Teacher's Guide*. A school project creating a regional treasure hunt to help community members discover special places. Keene, NH: Antioch New England Graduate School, 1996.

*Young Geographers*, by Lucy Sprague Mitchell. One of the classics in the field. New York: Teacher's College Press, 1934. Rpt. New York: Basic Books, 1963.

## MAKING MAPS

In the earlier years, children tend to represent their spatial environments through *percepts*, namely, using their imagination to draw or otherwise represent their world in free form and largely unaware of patterns or reference points—what we might term the mathematics of map construction. By the age of ten or so, the logic inherent in mapping begins to emerge, and so it is at this point we can begin to draw on these growing abilities to foster the learning-to-read process.

There are various ways of going about this, of course, but basic to it all is continuing the opportunity to write (create) maps from personal experience. The reader should *not* put aside the notion that, ultimately, map reading, like politics, is local; the most meaningful uses are found in maps that represent familiar areas or those for which the reader has a potentially strong personal interest. Geographers seem to forget this when they create maps for instructional use. The result is a plethora of pull-down maps showing huge areas of the earth's surface; no wonder we become distracted by rudimentary bits of information about location and other specific bits of facts. Attempts to build usable concepts founder when we use examples that are highly generalized character. We should, rather, look for opportunities to get more of a handle on the basic concepts of mapping: how ideas of location, direction, and distance can be represented in a precise way. There are many ways this process can begin. One, described by Greenhood (1964, pp. 206–210), begins with what he calls "footmade maps." Although even here, many subtleties may be introduced, the basic idea is simple enough: employing one's own pair of legs to pace off distances and directions on a familiar piece of turf. Students can start with the lot on which their house sits, all or a portion of a nearby park, the more school ground, or any bounded area. The process can become more complex if there is introduced the problem of averaging the length of a student's "pace" and the adjustments necessary if the area being mapped has some slope to it. If the plot is marked off at right angles, then the question of how to create a 90° angle arises: the solution requires no more than a piece of string tied to a pencil or other marker and a straight line from which two sets of arcs can be drawn, denoting the line at right angles to the first—which students can, and should be asked to, figure out mainly by themselves. If the four sides are not of equal length, then the method is one of *trilateration*, in which the area is measured as two triangles (the long side of the triangle is the measurement from the points catercorner to one another). If the sides of a triangle are known, then it is not necessary to know their angles in order to construct other (smaller in this case) triangles representative of the larger one. Once the perimeters of an area are found, filling in the details by recording the location of items within their bounds, begins, which again is accomplished by pacing off distances from corners, midpoints, or other obvious markers.

When rendering a large area to a much smaller one (usually, but not always, the case in map construction), the problem of scale enters the picture. In the beginning stages, students can experiment with what happens in expanding or reducing the scale as one moves from actual areas to smaller ones or the other way around, or as attempts are made to find the appropriate size for a finished product. The principle to be discovered is that doubling the scale will quadruple the area (with the obverse operating as well). Of course, there may be occasions when it is desirable to enlarge or reduce a map to use it for other purposes. One instance is the case where a class for one reason or another wishes to create a very large version of the continental United States or other large portion of the earth on the school yard or other large area. To accomplish this, four right angles will be needed to mark a perimeter for the map. Students will also need to experiment with scales so that in transferring the map to the larger surface, it will be made the correct size for transfer of the various borders and symbols they wish to enter.

Footmade maps are created without reference to directions as such. In the *compass traverse method*, directions are taken with a compass and distances measured by pacing (at this point, each student has measured his or her stride a number of times in order to find an average for various types of terrain). It is, of course, advisable at first to select terrain that is fairly hospitable to the method: avoiding steel structures that could influence the compass reading unduly and finding an area that is sufficiently open and free of difficult obstacles. This method allows for mapping larger areas, but it also introduces some of the problems of taking accurate compass readings.

One more step up the ladder of accuracy in mapping is the *plane table method*, in which students are introduced to the basic concepts of surveying. As in the previous instance, there are several gradations of complexity. In the beginning stages, students can get by with commonly available materials: compass, tripod, spirit level, a "plane" table or large board, pocket knife, pencil, and sharpening material. More advanced/accurate stages involve specialized equipment such as a professional surveyor might use (alidade, theodolite). However, mapping with considerable accuracy, and certainly much satisfaction, can be accomplished with knowledge that hardly goes beyond high school geometry.*

## IN SUM

Children in British infant schools begin each day by writing a weather observation in their composition books. Four-and five-year-olds keep it simple with:

*David Greenhood, in a 1964 book titled *Mapping* (which is unfortunately now out of print), presents a detailed, yet clear, discussion of these three "methods" of creating maps.

Today is [Wednesday].
It is [cold, warm, cloudy, etc.].
The date is [May 17].

By age six or seven, the weather "story" has expanded:

Today is [day and date].
It is a [sunny, cloudy, cold, etc.] day.
The temperature inside is [degrees Celsius].
The temperature outside is [degrees Celsius].
It is a [south, west, etc.] wind.
The wind is a force [Beaufort scale reading].
Yesterday was [cloudy, sunny, etc.].
Tomorrow will be [cloudy, etc.].

Daily observations, of which this is but one type or example, heighten awareness of the natural world and show how that world impinges and interacts with human activities of various kinds. Children can be oblivious to such matters unless someone—a parent, a teacher—calls attention to them in a fashion with some inherent interest to it. Additionally, of course, we see in the example above how (print) reading and writing become integrated into what might also be called an aspect of the geographic curriculum.

Research in the development of cognitive, or thinking, abilities suggests how primary is the need to emphasize personal observation and interaction with the environment. When systematic attention to the real world is lacking, it is not only doubtful, but probably impossible, for a person to attend to maps and textbooks intelligently, since they are largely based on abstractions in the first place. Intelligent map reading would appear to grow most fruitfully from a background of experience, which includes, beyond the firsthand experiencing of the environment, many opportunities to represent those experiences. Just as one learns to read by reading, abilities in writing (here referring to the preparation of maps) grow best where writing is practiced rather than studied in isolation from purpose. As well, just as we know that writing plays a highly significant role in strengthening and expanding (print) reading abilities, the writing or creating of such representations will strengthen the ability to read the writing (or maps) created by others. The most direct route to fluency in map reading, therefore, appears to be the gradual, but tempered, introduction of the technical principles lying behind map construction.

## REFERENCES

Allen, Roach V. 1961. The language-experience approach to reading. In M. P. Douglass, ed., *Claremont Reading Conference, 25th Yearbook*. Claremont, CA: Claremont College Graduate School, pp. 59–66.

Deverson, H. V. 1948. *The map that came to life*. New York: Oxford University Press.

Edwards, Betty. 1979. *Drawing on the right side of the brain: A course in enhancing creativity and artistic confidence*. Los Angeles: J. P. Tarcher.

Efron, Robert. 1990. *The decline and fall of hemispheric specialization*. Hillsdale, NJ: Lawrence Erlbaum Associates.

Greenhood, David. 1964. *Mapping*. Chicago: University of Chicago Press.

Harris, Lauren J. 1985. Teaching the right brain: Historical perspective on a contemporary educational fad. In Catherine T. Best, ed., *Hemispheric function and collaboration in the child*. Orlando, FL: Academic Press, pp. 231–275.

Hunter, Madeline. 1976. Right-brained kids in left-brained schools. *Today's Education*, 65, 45–48.

Makower, Joel, ed. 1992. *The map catalog: Every kind of map and chart on earth and even some above it*. 3rd ed. New York: Vintage Books.

Markham, Beryl. 1942. *West with the night*. Boston: Houghton Mifflin.

# 5

# Toward a Geography of Geography in the Curriculum

Dear Abby: When a letter comes back from the post office twice with a notation informing the sender that mail to Kansas requires Canadian postage, it is not merely a ''gaffe'' but a symptom of a more serious problem—national ignorance of geography and history.

As an international trade lawyer, I have had letters to Hong Kong returned after being routed to Guatemala.

Telephone operators have asked me what state Ottawa is in, where in China Singapore is, and in which part of Yugoslavia Dusseldorf is located.

When postal workers think Kansas is in Canada and Hong Kong is in Guatemala, it's time to take a serious look at education in this country.

M.A.C. in D.C.

Dear M.A.C.: You're singing my song.

*Los Angeles Times*, February 2, 1993

Michael Jordan was a geography major at North Carolina. Said Jordan: ''I knew that I would be going places, and I just wanted to know where I was when I got there.''

*Los Angeles Times*, November 12, 1992

## WHERE IS GEOGRAPHY?

Potentially, since we are surrounded by and live within a spatial environment, geography is everywhere. However, as human activity increasingly alters and makes more artificial this ''home of man,'' the underlying reality of the physical and natural world becomes more difficulty to see, understand, and ultimately appreciate. This transformation, as represented by the asphalt jungles of urban areas, results in a world that, for all practical purposes, obscures the natural environment almost completely. Even in the typical suburb—lest we need a reminder that the vast majority of America's students live either in its cities or their suburbs?—what nature has provided has been altered, often dramatically, and in the process, has often been sorely misused. Nor have those areas set aside as open space within these highly developed areas—most notably, the public parks and playgrounds—escaped (see, for example, Kunstler's 1993 critique of the so-called urban jungle, *The Geography of Nowhere: The Rise and Decline of America's Man-made Landscape*).

Particularly where the schools are concerned, trees and other vegetation have been uprooted and hauled away to provide open areas, the land leveled, and much of it then covered with hard surface and, perhaps, some grass, all to accommodate a curriculum of organized games. Public parks may not have received such drastic treatment, but most have also been graded and denuded of everything but a few trees and, perhaps, shrubs. More likely, the vegetation we do see has been imported from other environments, relocated according to someone's notion of a pleasing vista but with minimal regard for what was there before. There are, of course, exceptions, principally those that have been in place for a long time, having been created before the invention of machinery capable of totally altering the landscape with such precipitous and massive effect as those we have at present. As well, today's ''open spaces,'' its parks and other public lands, are not only relatively fewer in terms of the population they are supposed to serve, but fewer and farther between. Developers only yield to the demand for park land and other open space when its inclusion is stipulated as the condition for being granted permission to ply their trade. The result has been grudging acknowledgment and a minimal response to community needs for elbow room beyond the confines of a town lot or, worse, an apartment complex.

The ''natural environment'' is not all with which a geographer is concerned, of course. The study of geography in its mature form is concerned with all aspects of human behavior within the spatial environment. But when we consider how youngsters may become aware of, and develop an appreciation for, the environment in its broadest sense, it is important to realize that a significant portion of it—actually, the most important part—is not easily available for most youngsters to experience. *The Geography of Childhood: Why Children Need Wild Places* (Nabhan & Trimble, 1994), describes what the authors consider to be the kinds of circumstances and experiences that attract and hold the attention of young people and how these vary from the foci of adults, who often think that what interests them must also be of concern to the younger generation. For example, where adults look for the ''big picture'' (the attractive vista or panoramic shot for their camera), children fall on their hands and knees to examine the bits and pieces immediately before them. Those parts of the world that are important to them are composed of things to be directly seen, touched, smelled, and even tasted. It is also a world characterized by nestlike refuges; the authors point to studies that demonstrate a clear proclivity for microhabitats, ''secret'' places that screen children from the larger world and from which they can look out but where, they can easily convince themselves, no one can look in. This is not just the world of the preschooler. Young people, into their teen years at least, demonstrate this attraction to microhabitats. For example, one such hide-away plays a prominent role in a recent book that has become widely popular since its translation into some 30 languages. *Sophie's World: A Novel about the History of Philosophy* (Gaarder, 1994), written by a Norweigan high school teacher, tells the unlikely story of a thirteen-year-old's introduction to philosophical thought. A good deal of it is centered around a private place where Sophie goes to read secret messages about philosophy and to reflect on them. Tree houses, ''private'' signs on bedroom doors, clubhouses—even gang membership—provide other clues to the pervasive nature of this behavior.

It should also give us pause when we consider the layout of the typical American school playground; normally a wide open space with much if not all paved; anything remaining usually in grass or left to dirt; with, perhaps, a few trees scattered about and a shrub or two hugging the building's foundations. There are no places to hide, creating an interesting if disturbing contrast with the British ''adventure playground'' where, if there are also relatively few natural places to get away, various items—cast-off boards, pallets, and the like—provide opportunities for construction and the much-misunderstood privacy that young people crave. Although there are also plenty of opportunities for splinters and skinned knees, at least the youngsters can construct their own worlds and escape the realities that adults would like to enforce on them. I have commented elsewhere about the tendency of Americans to be safety-conscious and their inclination toward litigation, a proclivity that encourages the development of the sterile playground environments so common in U.S. schools.

That the playground often fails as a fruitful place for acquiring geographical ideas may also be true of the classroom itself. Gary Paul Nabhan writes, for example:

> From my first day of kindergarten to my first day of graduate studies in botany twenty years later, school was synonymous with staying indoors, out of touch with the most elemental aspects of life. Within my initial hour of formal education, I was yanked up by the shoulders and scolded for crawling behind a piano to curl up to sleep in a bookshelf. I had quickly come to the conclusion that Mrs. Wiltrout, the kindergarten teacher, was going to bore us simply by *talking* all morning long, so I thought I'd rest up for when she let us out for recess just before lunch.
>
> I had similar difficulty with the first day of graduate courses in botany two decades later. After a full summer of learning about desert plants from Indian people, I fell asleep in the front row of an air-conditioned lecture hall while being told that the Linnaean system was *the* only way to classify and study plants. I was simply better adapted to learning by doing, outdoors, than to learning by abstract formula.
>
> And yet, from Mrs. Wiltrout's first insistence on order in the classrom, to my uncompleted junior year before I left high school for good, I can recall only three school-sponsored field trips for nature study. We were allowed gym class and playground recess,

of course, but little time beyond the manicured confines of the school yard. During my first twelve years of school, I figure that my teachers offered a total of less than six uninterrupted hours in the marvelous natural laboratory at our doorstep: the Indiana Dunes, a hodgepodge of buried forests, quaking bogs, and mountains of sand. Even in a place so well-suited for nature study, my teachers kept us inside classrooms for a thousand hours for every one hour they took us into the field. In urban centers with no easy access to natural areas, I'm sure my contemporaries suffered even more severe sentences. (in Nabhan & Trimble, 1994, p. 38)

Gary Paul Nabhan nonetheless grew up to become a MacArthur fellow and a plant conservation biologist, but we should take to heart what he has to say about what students often meet up with in their classrooms. He also points out how students living in concentrated urban areas, as well as those who make their homes in the wide open spaces, increasingly report that their knowledge about the environment generally is dependent on vicarious sources, primarily upon television and motion pictures, but also on the increasingly ubiquitous CD-ROM. As we shall see in more detail in Chapter 6, vicarious experiencing, in which an artificiality of some sort is inserted between the learner and the thing itself, may embellish or enhance knowing, but it cannot stand alone as a source for knowledge, nor can it serve successfully as the starting place for learning.

## A WORD ABOUT TEACHING AND LEARNING

It is commonly said that school is a preparation for life, as if what went on in school was not really living. In one sense, it *is* a period of preparation, a time for acquiring what people generally believe will serve as a foundation for—perhaps first and foremost, although regrettably—becoming an economically viable member of society. But it also includes, or should include, a time when the basic attitudes and knowledge of an aware and participatory citizenry are developed.

Reflection reveals how inadequate is the "preparation for life" represented by school. Our own experience and that of our children tells us that school is living to them as much as our jobs are living to us. The problem is that the traditions that control much of what goes on in school in the name of teaching are themselves so artificial that what happens to students does, in fact, seem unreal, and so we think of it as a preparation for a reality lying outside the four walls of an institution we call *school.* When students enroll in that formally organized enterprise to begin with, we create "a cage for every age," as it were. From there we organize their instruction within this age-grade social structure, refining it as we go along by instituting within-and between-class groups based on how well the students master the curriculum. Within this frame, we teach a curriculum organized around "subjects," which are in fact figments of the mind. This is just as true for the elementary school as its subsequent counterparts, with the exception that young children are usually taught each of the so-called subjects by the same teacher.

The artificiality of this approach has long been acknowledged, not only within academe but also in terms of classroom practice. Scholars point to the increasing fuzziness in the boundaries that separate their fields of inquiry. The traditional distinctions have been breaking down commensurate with the knowledge explosion, which itself has been expanding exponentially for the last 100 years and has catapulted us into the Information Age. Geographers, particularly, should be aware of this phenomenon, given the fact it has spawned many subspecialties over the years.

Educationists from at least the time of John Dewey have urged the adoption of curriculums that reflect these realities. Unfortunately, however, they have not been much heard. Educational institutions are by nature resistant to change. Political problems are generated by the fact that education itself, and particularly the social studies (including, of course, geography) are, at their best, controversial, for these ''subjects'' involve values as well as specific information. Conflicting opinions about values give rise to the sort of debacle we witnessed on a grand scale with the MACOS project (cited in Chapter 1). It is much safer to avoid conflict, of course, and so there are great pressures to remove the human element, the thinking and valuing, from the classroom. In terms of practical, everyday teaching concerns, there are many pressures on teachers to keep things simple. They find it much safer to stick with the familiar and the seemingly successful, or at least what is reasonably successful where standard measures of achievement are concerned. Standard tests do not measure to any substantive degree the things that teachers who depart from conventional ways of teaching hope their students will learn. Disappointment at the result of so much energy spent without the so-called objective evidence that tests are said to provide can be demoralizing.

This is why the teaching of spatial ideas so often devolves into place geography, the mechanics of map reading, and the memorization of facts. These are not controversial topics. Nor are they inherently interesting, as they have no hook to engage students' minds. This is also why geography as a school ''subject'' shares with the rest of the social studies the reputation for being boring. Geography is not an end in itself; it is a way of thinking, literally a vehicle for thought. Approaching it as a means for extending understanding brings with it specific knowledge, the kind of knowledge that has a long and useful shelf life, but also that society values as an indicator of a successful educational enterprise.

Ultimately, teachers teach what they know, while the act of teaching itself is a form of revelation. If a person has any interest in finding out whether he or she has command over a set of ideas, perhaps the best way to find out is to try to teach about them. Every teacher has probably, at one time or another, attempted to teach something over which that command was unsure by ''staying a lesson ahead of the class.'' No one would seriously recommend this for any length of time. *Successful teaching depends on a depth of knowledge sufficient to free the teacher from the subject matter itself. Without such depth, both teacher and student become slaves to the irrelevant.*

## GEOGRAPHY, LIKE POLITICS, IS ULTIMATELY LOCAL

Like politics, geography is ultimately local. We can study about far-off places and doubtless gain from such inquiry. But its meaningfulness and the extent of its intellectual shelf life depend on the richness of firsthand experience. Jean Piaget, whose research remains fundamental to our understanding of how spatial understandings appear to emerge (presented in considerable detail Chapter 3), has provided a model for thinking about the thinking process generally. His theory of intellectual development was not developed de novo, of course. He was particularly indebted to John Dewey's (1859–1952) description of learning as the *reorganization of experience.* Piaget (1985) expanded on this idea, describing the learning process as having essentially three aspects, those of *accommodation, assimililation*, and *equilibration.* As in Dewey's view, every person, or learner, is in possession of what one might term an *experience base*: all the things that have happened to a given individual, knowingly or not, join together to form the foundation for the way that person perceives and thinks about the world of which he or she has a potential for being aware. We are, of course, more cognizant of experiences that we can think about directly, but feelings and attitudes—all the components of the world of *affect*—play a role in thinking.

Piaget's tripartite learning model conceives an ongoing, dynamic activity, in which new experience is mediated by the processes of equilibration: when the "new" meets the already extant experience base, there develops a need to find a fit for it. The new rarely, if ever, finds a place in the mind that was not previously colored or otherwise affected by prior occurrences. The resulting clash may be benign or extreme, but there is always the need to accommodate the new with what time has built up over the years (whether two or twenty), regardless of how few or many those may be. In Dewey's term, this activity (what we term "learning") is a reconstructive process; for Piaget, the term *equilibration* provides a more dynamic metaphor, highlighting it as ongoing and never completely at rest. In this model, learning is a continuous process, just as experiencing is one way to characterize living.

The relevance of any particular learning activity that a teacher might plan must, of course, remain within the experiential background of the student. We have known this at least since the time of John Amos Comenius (1592–1670), which may lead to some wonderment about why we might need to remind ourselves of this truism today. When we abandon allegiance to this principal, the resulting "understandings" may very easily be full of error. It is in this context that pink on a map means pink on the ground or where "up" on a map always means north. We do know, as well, that many students gain an intense dislike for school in general, and perhaps particularly social studies in its various permutations (including ''geography''), as a consequence of being taught things they perceive to be irrelevant (whether this is, in fact the case) or do not comprehend.

Piaget postulated that thought processes evolve over a period of years, proceeding in a series of stages from the very immature, egocentric behavior characteristic of the infant and very young child to mature thought, which, he believed, begins to emerge around the age of twelve but becomes consolidated only during adolescence. Piaget labeled this last stage *formal operations*, meaning that persons at this level are able to compare and contrast abstract ideas out of which a practical course of action might be anticipated, if not actually carried out. Relatively little research has been done on this developmental stage. However, if we stick with the kind of research protocol Piaget favored, there is considerable evidence that few adolescents can deal with such abstractions; and perhaps fewer adults than we might think are capable of such thought as well.

However, Piagetian revisionists, notably among them Margaret Donaldson (1978), have pointed out that it is very likely that children and young adults may be able to deal with abstractions much earlier on than Piaget envisaged. The caveat in this assumption is that circumstances must be much more ''user friendly'' than those in Piaget's research protocols. The conditions for learning must be arranged to become much more closely related, in the learner's mind, to situations which are real to them. In the Donaldson study, for example, very young children were able to put themselves in the place of another observer *provided* the situation in which they were placed corresponded closely to what they knew from their own experience. The lesson here, of course, is that our classrooms also need to be more user friendly.

The problem for geographic literacy is, therefore, uncovering how understandings can be elicited when the ''subject matter'' of concern cannot be experienced directly. We recognize this reality when we comprehend the meaning of the statement that geography is ultimately local. As educationists from Comenius onward have reminded us, one person's ''reality'' cannot be transferred directly to another; its construction depends on the learner's ability to make the linkage by calling on the only resource available: the personal experience base. I wish to emphasize the extended period of time it apparently takes to achieve some degree of maturity in dealing with spatial abstractions. I also want to underscore the need to focus on things that can be touched and ideas that can be manipulated within a concrete, rather than abstract, environment. We need to know more about the qualities that are foundational to theoretic thinking (Piaget's formal operations). It would appear in the meantime that one of the reasons why geographic ideas find little purchase is that students have, despite the good intentions of their teachers, been asked to deal with realities that they perceive as abstractions.

## SOME THOUGHTS ON GEOGRAPHY IN THE CURRICULUM

The usual response when the school is criticized for failing to bring its students to a presupposed level of competence in one or another area of the curriculum is to demand that more attention be given to its teaching. Invariably this also means isolating the offending subject matter for this purpose from the rest of the curriculum. The most obvious examples of this behavior are found in the reading and (to a lesser extent) arithmetic curriculums of the elementary school. No other country devotes as much instructional time to print reading as the United States. In Goodlad's study of American public elementary and secondary schools, for example, teachers in the primary grades reported that they spent almost half the school day on reading instruction (about one-fourth went for instruction in "mathematics") (1984, p. 199). Since teachers tend to underreport the time devoted to reading instruction (in an effort to conjure an image of a "balanced curriculum"), we may probably conclude that these figure are low. In contrast, schools in Scandinavia, where the problem-reader population does not exceed 3 percent, provide little formal instruction in reading beyond the first year in school, relying instead on literature and the content curriculum generally to carry the burden of print literacy development. While other countries that have reasonable aspirations of achieving universal literacy may have slightly higher levels of problem readers, none teaches reading so intensively or for so long as the United States—and none has anywhere near the rate of reading disability. One is led to wonder whether the intensification of instruction with the idea of improving performance may not be counterproductive. Exacerbating the situation is the demand that students who do not perform satisfactorily in elementary school be provided additional remedial instruction at subsequent levels, resulting in their being excluded from content oriented-classes while their problem is being "fixed."

Coupled with this uniquely American phenomenon are the cycles to which curriculum organizational patterns are historically subject. The recognition of the artificiality of the traditional subject matter areas has been around for some time, particularly at the college and university levels. This recognition has encouraged curriculum developers over the years to propose ways of breaking down the barriers that afflict the traditional curriculum at both the elementary and secondary levels. Various schemes have been suggested and numerous instances in which they have been put into practice have been recorded in the educational literature. For purposes of discussion, these can be identified under five different rubrics. The first three, Correlation, Fusion, and Broad Fields denote efforts to find ways of combining subjects without entirely losing the notion of the traditional discipline-centered curriculum. Beginning at the other end of things, away from the disciplines as the progenitors for setting the criteria in judging what should be learned, we find the fourth and fifth ideas, the Activity Curriculum and the Core Curriculum. Generally speaking, these are curriculum structures that begin with problems or activities, which then serve as catalysts for stimulating a search utilizing an array of sources appropriate to the solution of the problem or the satisfaction derived from completing a meaningful activity.

Historically, curriculum patterns have cycled between a tendency toward curriculum organized around different subject matters and one or another of the patterns that place process ahead of subject matter. The latter two, the core curriculum and activity curriculum were children of the Progressive movement and are now seen largely within those schools, mainly private, that are generally viewed as providing a curriculum derived from ultra-liberal thinking, and therefore acceptable to only a small fraction of the population. The view is commonly although erroneously held that these forms of curricular organization depreciate the importance of subject matter and that "real learning" is thus somehow compromised. The launching of *Sputnik* served as the catalyst in a period of growing distrust in the ability of the school to teach students, particularly in the sciences and mathematics. The result was a flurry of curriculum development efforts, initially in those fields, but subsequently in areas across the curriculum, and there was also a generally conceded failure where the geography curriculum was concerned. When these efforts failed to match achievement expectations, as in the instance, for example, of the so-called new math, there developed a reversion toward traditional teaching methods, which, not surprisingly, also failed to net the desired improvements, as measured primarily, if inaccurately or at least incompletely, by test score results.

Seesawing policies have resulted as the schools have responded to growing public criticism. Until relatively recently, the problem has been seen to involve the product/process dichotomy: should the school focus on the acquisition of subject matter as an end in itself or does the content curriculum serve a larger purpose, however generally defined? Now, however, the problem is beginning to be perceived as a structural one involving the total educational circumstance. The solution, many have come to think, lies in what has been termed *restructuring*, namely, creating a new organizational environment in which the school can carry out its business.

The idea of restructuring has taken a number of different forms, some primarily administrative and others intent on dealing with the curriculum itself. Human and monetary resources are being poured into the effort from both public and private segments of the society. Although such curriculum development efforts suggest this may likely be déjà vu all over again, there exists a major difference between this and previous efforts at school improvement. The earlier attempts were characterized by the creation of experimental curriculums touted to be on the cutting edge of new knowledge and advances in our understanding of the learning process. Contained therein was the assumption that, through the reeducation of a cadre of teachers for the new curriculum, there would be a gradual, almost osmotic, spreading of the new to replace the old.

The restructuring of the last decades of the twentieth century is much more distanced from actual classroom practice than in other periods. For example, proposals are made to reorganize large school districts into smaller units on the presumption that these ought to be more responsive to its constituents. In extreme cases, the British model of establishing autonomous school units freed of virtually all local and state regulation is being tried. In curriculum matters, still following the presumption that change can be brought about by administrative fiat, there have been major efforts to develop *national standards*. The presumption here is that such proclamations will bring some kind of order out of the curricular chaos that, most people think, not only characterizes teaching and learning in the United States but, if ended, will assure the excellence in learning that everyone desires.

Whether chaos is characteristic of the curriculum generally and geography in particular is open to question. In a 1990 book carrying the critical title, *The Predictable Failure of Educational Reform: Can We Change Course Before It's Too Late?* Seymour Sarason remarks on the widely held notion that the quality of education will improve if some kind of standard is invoked to control what goes on in the thousands of autonomous school districts around the country. Saying that the issue is ''akin to Charles de Gaulle's comment about governing France: 'How can you govern a country that makes five hundred different cheeses?' '' he asks:

> Is it . . . time to forge national criteria of excellence to which all school districts would or should aspire? Although it is understandable why such proposals get made—reflecting as they do a response to the intractability of the bulk of autonomous school districts to demonstrate improvement—they are examples of two things: missing the point and ignoring the obvious. The obvious they ignore is the point that John Goodlad (1984) makes in his heroic study of public schools: despite the many and obvious ways in which schools differ, they are amazingly similar in terms of classroom organization, atmosphere, and rationale for learning. The point they miss is that the classroom, and the school and school system generally, are not comprehensible unless you flush out the power relationships that inform and control the behavior of everyone in these settings. Ignore those relationships, leave unexamined their rationale, and the existing ''system'' will defeat efforts at reform. (pp. 6–7)

Despite such warnings and a U.S. Congress bent on reducing the influence of the national government, efforts to forge a national curriculum continue. Like the historians slightly before them, geographers and geographic educators have produced a document intended to do just that. Somewhat ominously titled *Geography for Life: National Standards 1994* (National Geographic Research and Exploration, 1994), it is a beautifully prepared publication representing the combined efforts of the American Geographical Society, the Association of American Geographers, the National Council for Geographic Education, and the National Geographic Society. Its subtitle, *What Every Young American Should Know and Be Able to Do in Geography*, suggests the intent: to redefine ''school geography'' in terms of concepts and generalizations that its authors believe should be taught throughout the elementary and secondary school. Eighteen ''standards'' and five ''skill sets'' are described and explicated in sets of criteria at three grade levels: kindergarten through grade four, grades five through eight (those now known popularly as the middle school years), and grades nine through twelve. The standards listed under each of these categories are comprehensive *and* intimidating, leaving as they do hardly any geographic stone unturned and, at the same time, appearing as a formidable barrier to translating their essence into classroom practice by teachers with little or no formal knowledge of academic geography.

Even this limitation does not negate the value in promulgating a set of ''standards'' as long as the validity of Sarason's (1991) caveats are held in mind. A very valuable consequence in publishing the *Geography Standards* is that it makes available to a broad and quite ignorant audience a comprehensive view of the nature of the modern ''discipline'' of geography. They will, I believe, be surprised to find how comprehensive today's *geography* is. This alone will help disspell the widely held misunderstanding that geography consists primarily of knowledge about places and things. The new geography standards may help the public see geography as a comprehensive and particular way of looking at the world and understanding its problems. The authors wisely have not presumed to tell teachers how these standards might be reached. However, because of the history of geographic education, geographers generally cling to the idea that standards can be neither realized nor appreciated unless geography is taught independently of other so-called subjects. As one geographer has written recently, ''[F]ew . . . would disagree with the proposition that geography as a separate subject taught through primary and secondary education would be an ideal background for life-long learning *and the integrative learning advocated in the university*'' (Enedy, 1993, p. 23; italics mine). Although accepting of the idea of integrating the traditional subject matters at the university level, Enedy rejects it for the earlier years, all the while acknowledging that it will probably be necessary to utilize what he terms a ''back door approach through other established subject areas . . . until an even stronger case is developed for graduation requirements that include geography'' (p. 23).

The responsibility for teaching geographic skills, concepts, and generalizations falls, at present, almost exclusively on the social science/humanities (i.e., the history) curriculum. It garners little direct attention from the natural sciences and mathematics. ''Back door'' strategists see the immediate future as one in which, of all the areas of the curriculum that might be enlisted for the purpose of teaching geographic concepts, special attention should be given to the role of the natural sciences and mathematics in this regard (Enedy, 1993, p. 23). This appears to constitute the main strategy for leapfrogging into the curriculum, and in this fashion making geography coequal with the other subjects.

I have documented in Chapter 1 the experience of geographic inquiry as it became subsumed within the so-called social studies, with the consequence that we have very largely overlooked the natural science and mathematical aspects of geography. But that problem is hardly likely to be solved by instating geography as a separate entity in the curriculum. It has long been evident that there are simply not enough hours in the school day to accommodate the many demands for equal representation. Put another way, the school day cannot be subdivided infinitely. But there is an even more cogent reason: the separate subject curriculum has already demonstrated its inefficiency as a pattern for educating students. Unhappily, the teaching profession has shown a considerable inability to think imaginatively about other ways of improving learning, preferring to intensify instruction within the existing framework rather than exploring alternative ways of engaging students' minds, and this in spite of a research record strongly suggesting that things taught in isolation bear poor intellectual fruit.

This is not to say that variations on the traditional patterns of organization have not been tried, nor that they have been found wanting. Many secondary schools, for example, at one time reorganized curriculums around something called *modular scheduling*. Computer technology makes possible the creation of class schedules of varying lengths and frequency of meeting, opening up the curriculum to many more options than are possible in the traditional five-, six-, or even seven-period day. Within this frame, combinations and specializations can be arranged for depending on specific and general objectives. Innovation raises a circular problem it is difficult to escape, however. Standardized tests do not measure the kinds of knowledge either educator Mark Hopkins (1802–1887) or Socrates sought to evoke in their students, and that constitute the more significant elements in what we generally call learning. Thus, when learning is measured only across a very narrow and restricted band or range, instruction tends to focus on this aspect of the curriculum. The result is an instructional program that slights the more important aspects in favor of producing the kind of evidence that the public has been taught to want. A more prosaic problem lies in the fact that modular scheduling changes not only the curriculum itself but the length of the school day for students and teachers. This means that students come and go at different times, not all at once as is traditionally the case. The public generally does not expect to drive by a school campus when the school is supposed to be in session and see students conversing or even coming and going. Passers-by may conclude that the students are not learning and that something is wrong.

Moreover, any movement toward change brings with it a degree of trauma, including a great deal of uncertainty among those who seek to institute it. Innovation always takes much harder work than it does to maintain the status quo. Then, when an aroused public is already criticizing existing practice at a high rate of intensity, perceiving at the same time that, in some distant past, school practices were somehow more effective, it is not surprising that school personnel become hesitant to break away from traditional molds, no matter how professionally obvious the need may be.

The "back door" metaphor is apt, at least as it suggests a transitional strategy toward creating a geographic room of its own within the overall curriculum; it also draws attention to the precarious place of geographic ideas in the school curriculum and gives one reason to focus particular attention on its special plight. With the formulation of the idea of the ''social studies'' early in the twentieth century, a problem arose for which a solution has never been satisfactorily achieved. While the idea of *process* remained imbedded in it, content could be derived from any and every relevant source, allowing the kind of integration of ''subject'' that we now increasingly see in postsecondary education curriculums. As that gave way to a concentration on *product* and the acquisition of subject matter gained ascendency over *learning how to learn* by utilizing multiple sources of information as a coequal goal, the essential contribution of the natural sciences and mathematics, particularly, dropped by the wayside. Geography, when attended to, became subsumed within a ''social studies'' curriculum that emphasized (when it was given any attention at all) the specifics of location and the characteristics of place, concepts that cannot in any intellectually honest way stand alone without considerations that draw on understandings deriving from the natural sciences and mathematics.

## "GEOGRAPHY" ACROSS THE CURRICULUM

### The Special Case of Science and Mathematics

Geographers may bemoan the state of their ''subject'' in the school curriculum, but this problem is not unique to them. Science education is also in exceptionally dire, and perhaps even more drastic, straits. Science teachers, like geography teachers, are unprepared, especially at the elementary level. Students also have a negative attitude for which, at least for the time being, there appears little chance of solution. Science education, again documented by Goodlad's (1984) research, shows the science curriculum garnering even less instructional time than the social studies, a position it maintains throughout all the school years. While this suggests a problem of balance in the curriculum, there is a contingent problem where science education is concerned in the deep dislike students develop for this ''subject,'' a phenomenon that emerges with considerable force during the junior high school years (Dimit, 1989; Eichinger, 1990), ironically just when science fairs and the like are planned with the specific purpose of capturing the long-term interests of students at a point often considered to be critical in the career development process. The cause for this negative attitude (a problem from which students do not appear to recover in high school) is something of a mystery since elementary students exhibit positive attitudes in this regard, despite the fact that instruction is meager and often sporadic.

Mathematics faces a parallel problem. Here the issue is not so much a matter of interest but of understanding. However, mathematical literacy takes second place (but barely) only to verbal literacy, especially print reading, as a desired outcome of schooling. But here I speak mainly of the most fundamental kinds of arithmetical calculation, which are seen as foundational; the rest, it is mistakenly presumed, will take care of itself if there is mastery of the four ''basic functions'': addition, subtraction, multiplication, and division. No curriculum area has been as subject to the pendulum swings incited by the product/process conflict as the arithmetical/mathematical. Oddly enough, given the conflicts that have arisen over what should constitute mathematics instruction, the introduction of curriculums designed to enhance understanding (process) over skill (product) has originated for the most part from mathematicians and mathematics educators who assert that the *process* in experiencing numbers (Spencer & Brydegaard, 1952, p. 4) is the key to developing mathematical understandings.

The ''new math'' of the 1960s, which constituted the first full-blown entry of math specialists into curriculum development in this field, came about as a result of a widely held concern that the usual method of teaching, relying almost exclusively on memorization, was ineffective in preparing a society to compete with the Soviet Union after the launching of *Sputnik*. Inadequate teacher backgrounds, along with parental mathematical illiteracy and the feelings of inadequacy flowing therefrom, caused a backlash that destroyed this particular effort at curriculum reform. Now, a generation later, the matter is being replayed. After approximately twenty years, during which traditional methods have dominated instruction (not that the new math ever gained the ascendency critics claimed for it), rote learning has come under fire and sporadic attempts have been made to introduce understanding into math. The critics, not waiting for any large-scale reintroduction of the process model, are once again attacking. It is the usual case in which something unknown is perceived to pose a danger that must be attacked before it gets out of control. The problem in all this remains much the same: parents have a conception about how knowledge of mathematics is acquired, based largely on their own perception of how they think they learned what they know, which then becomes the standard by which their children should learn. Teachers are as likely to share this view with parents, with whom they have perhaps more in common than with professional mathematicians and math educators.

The problem of the new math is largely an elementary school phenomenon, but it affects every subsequent instructional level. Not surprisingly, the problems at this level persist into the secondary school and even beyond. One indicator of its pervasiveness is the rapid growth of college level remedial courses where, as is the case with verbal literacy, entering students are increasingly failing to demonstrate a sufficient command of mathematical concepts on their entrance examinations.

Saying this, it is still difficult to perceive of any problem or issue involving spatial concepts that does not also include ideas having their roots in either the natural sciences or mathematics, or both. From the simplest kind of geographic ''problems'' very young children might encounter—counting and measuring phenomena, sensing direction, following weather patterns—to quite sophisticated studies of areal phenomena and relationships, concepts from the natural sciences and mathematics are necessarily part of the geographic equation. Geographic exploration provides particularly cogent opportunities to discover mathematical (or physical science) ideas rather than receiving them as revealed truth. For example, discovering the principles of right angles and differentiating scale becomes a natural outcome when students seek a solution to another problem: attempting a maximum degree of accuracy in reproducing, and in the process enlarging or reducing, an existing map. The teacher might, of course, adjust the scale to fit the assigned space (perhaps an area on the playground where a map depicting the original thirteen colonies is to be drawn) and produce a compass or protractor with instructions in its use. Alternately, the students might be supplied with a piece of string and, by attaching it to a pencil, experiment with finding how this simple device may be used to construct a right angle, the basis of the grid system. In a similar method of hypothesizing, experimentation, and validation (i.e., whether the final product "fits" the space selected), an understanding of the concepts of scale and ratios may be derived in the same spirit of discovery.

The ubiquitous nature of such opportunities is such that one might easily wonder why more is not made of them. I can only suggest that we suffer from a form of tunnel vision when a certain "subject" (apparently, most notably mathematics and natural sciences) needs to be taught, thus lessening the chance for the kind of subject matter integration that is implied.

## The Spatial World through Creative Literature

I use the term *creative literature* here to distinguish those things written to delight or edify rather than materials designed primarily to be instructive. We have far too much of the latter in our schools, and too few of the former. The slavish adherence to curriculums designed around the almighty textbook surely lends credence to why "school daze" is such a descriptive term. A major problem lies in the language of textbooks, which for the most part are written in a largely incomprehensible vocabulary—"field-testing" instructional materials ahead of publication to see how effective they are in helping students learn has never interested publishers. Textbooks and teachers' dependence on them are, in large part, responsible for the boredom that is endemic in our classrooms, and that flows from the inability of many students to make sense of what they are being told to read.

Although it often appears that the only "literature" with which students come in contact is in school, that is not the case. Somehow, our society does produce a large number of avid readers. Library circulation figures show an increase in circulation that more than matches the growth in population, despite decreases in funding for both public and school libraries. Nonetheless, unfortunately, very large numbers of students are not out-of-school readers and many learn to dislike the idea of reading so much they will derive no pleasure or edification from the activity and will simply avoid it.

What we might include under the rubric of "literature" is obviously virtually unlimited in its scope. It incorporates what might be called imaginative as well as critical writing, fiction and nonfiction. Even the most restrictive and/or traditional curriculum includes works of both genres, as present primarily in the English curriculum of the secondary school (an outstanding example is found in Samuel Clemens's *Life on the Mississippi*) and the so-called reading curriculum in the elementary school (see, for example, Miller, 1989). However, the fiction and nonfiction available to teachers of all subjects provide many opportunities for stimulating students to think about ideas that are ''geographic'' in nature.

It may seem surprising that geographers have for a very long time been interested in the contribution that literature can make toward geographic literacy. As early as 1924, for example, J. K. Wright said:

> Some men of letters are endowed with a highly developed geographical instinct. As writers, they have trained themselves to visualize even more clearly than the professional geographer those regional elements of the earth's surface most significant to the general run of humanity. (Quoted in Salter & Lloyd, 1976, p. 1)

Salter and Loyd go on to say that:

> The search for meaning and order in the landscape—that is, the desire to see landscape more clearly and completely—is a primary concern of geography. This search leads to landscape description that looks beyond the more obvious forms and functions into the deeper human implications of the world around us. When we apply creative writing to support our geographic vision, we have gained a powerful ally toward our goal of communicating a sensitive, articulate image of the phenomenon we call landscape. Creative authors intentionally use landscape to convey meaning within the context of a story and its characters. This rich interaction among the various parts of a creative work produces a landscape sense which excites the geographical imagination. This deeper insight can in turn lead to a more creative geographical description of landscape actuality and potential. (1976, p. 2)

Emphasis needs to be placed on what the authors term ''the geographical imagination,'' since fiction, at least, is open to a range of interpretation. Other forms of literature may still edify while being more objective in their descriptions. Charles Pellegrino's book, *Unearthing Atlantis: An Archaeological Odessey* (1991), for example, provides a fascinating and vivid description of the eruption of Mt. Pelée on Martinique in 1902 and of the vast changes that occurred in the landscape following this humongous disaster:

> At 8:02, as a knot of about forty people stood on a hill overlooking the city, preparing to descend for the morning mass, someone in the telegraph office tapped out ALLEZ—the last word anyone heard from St. Pierre.
>
> Then, where the mountain had been, a ball of glowing red dust appeared. It swelled, tripling its size in a second, then tripling again. At its edges surged through cool air,

the ball faded from red to black. Four miles away, amid the hillside knot of people, Fernand Clerc stood with his family. They held each other close, waiting for death and wanting to die together. The thing held Fernand spellbound—there was something dreadfully beautiful about it—all roiling black hell and expanding dimensions. For a moment it seemed fixed to the mountain; but then part of the cloud broke off and, with astonishing speed, advanced along the ground toward the city.

The soil and rocks rumbled, but there was as yet no sound in the air above. Shock waves are transmitted more rapidly through the solid earth, so the Clercs saw the top of the mountain burst apart, and felt the release of some twenty or thirty kilotons beneath their feet, eight seconds before they actually heard it. And when they did hear it, it arrived as a concussion of trapped air strong enough to knock them down.

''Above, a column of dust mushroomed and blotted out the sun,'' observed Mrs. Clerc. ''Below us, we saw a sea of fire cutting through the billowing black smoke and washing over the three valleys that were supposed to protect the city from Pelée. And incredibly, within these high-walled valleys, there existed havens of refuge where trees and bushes survived in the lee of one massif—the plane of destruction passing harmlessly overhead. The cloud rumbled over and over. One moment it would clutch at the ground, the next it would rise perhaps a hundred feet before falling back to the earth again. It seemed to be a living thing. It leapt over the botanical gardens, and in places even doubled back on itself, traveling the way it had come.'' (pp. 67–68)

Becoming aware of the landscape and of spatial interactions generally is not confined to more mature forms of literature. The now very large number of books classified as children's literature provides the widest possible range of resources; geography is literally everywhere in books written primarily for children (see, for example, Oden, 1992).

Many years ago, William S. Gray, ''Mr. Reading'' in the United States for over fifty years, pointed to what he presumed to be a fact beyond challenge, that every teacher is a teacher of reading (and hence, of literature). If we take this admonition seriously, then we must also realize that every teacher of ''geography'' also has the responsibility of recognizing and acting upon opportunities to extend students' literacy skills and appreciation through this strand of the curriculum as well, a point illustrated by a group of British writers collected in a volume bearing the title, *Language and Learning in the Teaching of Geography* (Slater, 1989). As in so many instances, the tendency toward compartmentalization of responsibility where teaching is concerned means that we are inclined to leave such instruction to that portion of the elementary school curriculum devoted directly to print reading, and for older students, to the teacher of English. It should be of more than passing interest to Americans to appreciate the fact that in Britain, where ''geography'' is a long-established subject at virtually every instructional level, there is such a strong interest in assuming responsibility for developing literacy in its more conventional sense within the framework of a long-established curriculum tradition.

## The Arts and Geography

As in reading and writing in their traditional sense of dealing with oral language and its written (or printed) form, we both create, and are observers or readers of, the arts. Here I speak primarily of art in its representational sense, and not so much performance art, although geographic references are also found throughout this area of the arts. For example, in the case of music, there are many opportunities to heighten awareness of spatial or geographic phenonena. Ideas can be generated from traditional and folk songs (*Home on the Range, On Top of Old Smokey, Red River Valley*), music in a classical vein (Edward MacDowell's *Woodland Sketches*, Modest Mussorgoky's *Night on Bare Mountain*), and perhaps even especially in recent popular songs. For example, Byklum (1994) argues that contemporary music employing spatial/geographic settings provides, not only a source for studying popular culture, but a way to match the interests of students with their daily lives.

Paintings and films provide a wide range of opportunities for heightening awareness and broadening perspectives. In a world increasingly dependent on visualization in communication, the number of such resources appears close to infinite. It is consequently not so much a matter of what might be available, but how a particular medium might be woven into the curriculum. The operative component here is the idea of integration rather than the didacticism that we typically associate with classroom practice. Not a few adults remember teachers whose approach to instruction involved such minute dissections that they emerged with an intense disliking for a painting, particular piece of music, or other artistic icon, a result that is certainly unfortunate. A certain subtlety or nondirection is, therefore, doubtless advisable. The most important idea the arts can communicate to students is the ubiquitous presence of the phenomenon of space as an integrating concept in thinking generally.

Saying that, we all can find an infinite amount of detail in a Gifford landscape or the work of other nineteenth-century artists, particularly those of the Luminists or the Henry Hudson School. More contemporaneously, we might think of the photographic works of Alfred Steiglitz Edward Weston, or Eliot Porter (see, for example, those appearing in Porter's 1962 book, *''In Wildness Is the Preservation of the World''*). One is of course not confined to American artists; painters and photographers have recorded images of all sorts that enlighten succeeding generations in all kinds of ways, among them the geographic.

## In Sports: The Case of Orienteering

Orienteering is both a highly competitive sport and a source of physical activity, largely without the rivalry we associate with organized competitions. In its simplist form it may not even employ the use of the compass and contour map, which typify the sport in its usual form. However, it does usually require some legwork (although variations have been developed so the physically handicapped may engage in it as well). All that is required is to mark out several destinations over a course that returns to the starting point, and you're on your way. It is an activity available to virtually any age group and can be carried out in a classroom, on a not-too-small playground or park, in larger park lands, and in open country (see, for example, Kjellstrom, 1976; McNeill, Ramsden, & Renfrew, 1987).

Orienteering has long been a popular sport in Europe, (especially Scandinavia,) and the Orient, having its origins in the Norwegian and Swedish armies some 100 years ago. It is expanding in the United States under the auspices of the U.S. Orienteering Federation, whose membership exceeds 7,000 members enrolled in some 65 clubs throughout the United States. Similar orienteering federations are now active in 40 countries. It is less a spectator sport than cross-country running, however. There is simply no venue from which an audience may view the proceedings except at the finish line. Orienteers, as participants are called, ply their game very largely alone or in pairs and out of sight, spreading out over a large area in pursuit of their landmark targets and appearing in the end only after they have punched the distinctive mark at each of the predetermined control points on a card to prove the course has been completed. Orienteering is not only a summer sport; events are also held during the winter on cross-country skis. Other approaches include orienteering on mountain bikes, horseback, and in canoes. In urban environments, orienteering events even utilize mass transit systems and local shopping malls as a part of the course. Although it can become a highly competitive sport—it draws literally thousands of participants, even in such a small country as Norway—the concept of orienteering is adaptable to all age levels (Conniff, 1992).

In their comprehensive discussion of the availability of orienteering for virtually any age group, McNeill and colleagues (1987) provide an outline of activities and objectives applicable for students from the primary years through adulthood (see Table 5.1).

The idea of geography in sports is a relatively new phenomenon which will develop in the future as more resources become available

## Community Resources

Every community, even the most urbanized, provides an abundance of possibilities for the development of geographic concepts and generalizations. Consider, for example, this description of a Boston city school that has involved its students in a wide range of community research activities:

> To the north a student focuses on planes from Switzerland landing at Logan Airport. To the east a student observes container ships from Europe and liquified natural gas tankers from the Persian Gulf entering Boston Harbor. To the south a student can see traffic heading south to Providence, Rhode Island, New York, and ultimately, Florida on Route

**Table 5.1**
**Orienteering Activities and Objectives**

## Age 7-10 years

*AIMS*
* *to create an interest in the sport of orienteering.*
* *to develop an understanding of the map as a picture of the ground.*

MAPS
◊ desk maps
◊ classroom/hall
◊ playground (large scale and picture maps)
◊ park (1:5000, 1:10000)

SKILLS - what to teach
• plans, shapes - map as a picture
• colours, line features, buildings
• orientation (setting) of map to terrain

SKILLS (cont.)
• following easy line features - paths, tracks, progressing to streams, walls, fences, fields
• no route choice - controls on line features initially, at every decision point
• map contact - folding, thumbing
• control points - punching, codes
• control card and control descriptions

TEACHING METHODS
Δ map walks with instructor
Δ string courses, picture maps, direct method
Δ 'white' - easy 'yellow' standard (controls at every decision point), pairs or individual
Δ cross country and score orienteering
Δ star exercises

EVALUATION
Games, worksheets, successful completion of orienteering courses

## Age 9-12 years

*AIMS*
- * *to develop an interest in orienteering*
- * *to develop map reading skills as an aid to navigation*
- * *to introduce the forest as one of several outdoor environments*
- * *to develop self-confidence through successful decision-making*

MAPS
- ◊ classroom/hall
- ◊ playground, centre, camp
- ◊ park
- ◊ woodland (1:5000, 1:10000)

SKILLS
- map legend and scale
- orientation of map to terrain, introduce the map-guide compass
- following line features, use as handrails
- recognition of features beside lines, eg. crags, boulders, buildings
- one obvious route, short cuts to path routes
- introduction to contours (hills, steep/flat)
- map contact - folding, thumbing
- organisation of control card and descriptions

TEACHING METHODS
- Δ 'white'-'yellow' courses, plenty of controls, on line features
- Δ pairs or individual
- Δ cross country, score and line-o, star exercises
- Δ badge incentive scheme

PHYSICAL ASPECTS
- + fun runs, games, circuits, relays with map

EVALUATION
Games, worksheets, successful completion of orienteering courses

**Table 5.1 (continued)**

**Age 13-15, 16-17 years and adults**

*AIMS*
*To navigate through the terrain with the aid of map and compass*

MAPS
◊ (classroom/hall)
◊ playground, centre, camp
◊ park (1:5000, 1:10000)
◊ woodland, forest (1:10000, 1:15000)
◊ mountain maps (1:25000, 1:40000, 1:50000)

SKILLS - INTRODUCTION (A)
- map legend, scale
- orientation of map to terrain
- use of line features as handrails
- simple route choice
- rough 'O' - long legs with good catching features
- fine 'O' - short legs of detailed map reading
- use of attack points, aiming off with compass
- basic use of compass/thumb compass
- distance judgement
- development of basic contour appreciation

TEACHING METHODS (A)
Δ 'yellow' - 'orange' - 'red' courses, as individuals
Δ colour coded courses

EVALUATION (A)
Games, puzzle-O, worksheets, quizzes, colour coded badge scheme

SKILLS - DEVELOPMENT (B) (see coaching handbook)
- building of contour perception into navigational techniques
- orienteer along large hills, distinct marshes, clearings, thickets; use of vegetation changes
- read contours in detail
- further use of compass and pacing
- simplification of routes
- race preparation and event analysis

TEACHING METHODS (B)
Δ 'green' - 'blue' standard courses
Δ self-programmed project (see 16+ modules)
Δ lessons and exercises in 'technique training'
Δ badge incentive schemes

PHYSICAL ASPECTS (B)
+ training for orienteering as a running sport
+ warming up and stretching (especially adults)

EVALUATION (B)
Progress through colour coded courses

*Source*: McNeill et al. (1987). Reproduced with the permission of Harveys.

95. The theme of movement in geography is easily taught when student and teacher share lunch on a six hundred foot high hill overlooking Boston. (Daly, 1990, p. 153)

Daly describes how the sixth through eighth grades in a Boston middle school have been organized to explore the environment through direct experiencing. Nearby is Thoreau's Walden Pond; adjacent are the physical settings of the New England coast and the local Blue Hills, where students hike and study the impact of population growth on the harbor below. Older students make trips farther away, for example, to the John F. Kennedy Library and to Washington, D.C.

A contrasting example, involving a rural community and younger children (first and second graders) is described by Guitierrez and Sanchez (1993). In a rural community located in the foothills of the Sangre de Cristo range of the southern Rockies of New Mexico, these teachers have established an outdoor learning laboratory on a hilltop near their school. As they become more aware of their environment, these students first laid out a ''compass rose'' (much as Henry the Navigator did in the sixteenth century on the southwestern tip of Portugal) and began recording observations: weather and climatic changes, plant growth, changes brought about by human activity, and so forth.

Students of every age will find a variety of possibilities for thinking about the environment simply by looking out the window, and even more when the concerns of the classroom take on a venue larger than the textbook or other assignments that intentionally restrict their view to things immediately in front of them.

Every culture and subculture finds expression within the landscape: in the type and configuration of houses and buildings, in the kinds of businesses and services that typify a community, in responses to regional climates (including the kinds of plantings, as well as other responses to the physical circumstances, including the location of arteries of communication), in the kinds of cultural activities, and so on. Direct observation becomes focused through the use of various resources, including telephone and business directories, cemetery records, local and regional statistics available from governmental and private agencies. More specific studies of, for example, traffic patterns involving both public and private transportation reveal information about a community that may have uses beyond the interest generated by the research activity itself. In one instance, for example, a fifth grade class in Oregon made a study of the ways traffic patterns were being altered in an area where various changes in the roadway system were being made to accommodate the construction of new public buildings and the development of a park, all of which resulted in a successful call by the students themselves for the installation of a traffic light (Zirschky, 1989).

*Historical geography* is a term applied to the study of the spatial characteristics of an area over time. City and country records provide a rich source of such data. On-site observations reveal the extent and nature of changes that have occurred in a given place following the collection of the initial historical data. Hardwick (1990) describes a class research project of this type conducted in the city of Sacramento, California. Utilizing city directories published since 1850, students were able to identify land use changes in ethnic patterns throughout the older portions of the city. While conceding that the primary means of identifying ethnicity is marred by the inability to account for some groups (especially in this case blacks and married women), the data provided a reasonably accurate picture of several different groups, whether they settled in concentrated or dispersed areas, the kinds of businesses in which they engaged, and evidence of their social and cultural activities such as church and club affiliations.

City directories provide another source for studying a community's history within a spatial context. City and county records include maps, verbatim accounts of planning and other stories of development as well as discord; local newspapers provide a rich resource for both written records and pictures. Oral histories provide contexts for understanding changing environmental and land use issues. Studies of particular natural events have been recorded and remembered (floods, earthquakes, fires, volcanic eruptions, violent storms), providing vivid illustrations for research and contemplation.

The American Southwest is particularly adaptable to studies of the effect of wind and water erosion as a modifier of the environment. One such study of an alluvial fan, a land formation created at the base of a mountain range where a steep slope combined with periodic heavy rains results in the flow of alluvium onto the plain below, has been carried out by elementary school students in a southern California community. The students begin by noticing how the land sloped away from their school grounds and how a seemingly level bike ride from the home to school was not as level as it first seemed. From there, field trips to collect soil samples from the top of the fan to its base revealed changes in its composition and consequent differences in vegetative growth. In this area, temperature gradients and the presence of differences in types of citrus trees (which are subject to frost at different temperatures) provide vivid examples of the principle that colder air moves toward the lower elevations, making frost more likely at the lower points of the fan. Experiments with a sluice box tell why the soil samples collected at the bottom of the fan are made up of finer particles than those farther up the slope. Students thus become aware of the various ways human activity is directed toward conserving water: through percolation basins, and, in this semiarid area, by pumping. Materials used in house construction suggest how geography influences human communities. Mapping, writing about experiences and experiments, constructing exhibits, and drawing illustrations help students fix the study in their experience. Every community offers possible variations on this theme.

Even in areas where field trips are difficult to arrange there is always the immediate neighborhood. Many resources may be found within walking distance of most schools. But it is not only the destination that calls for attention. Teaching the power of observation means encouraging students to become aware of the environment. Mapping those areas formalizes the situation. There are, however, other ways to help students raise their levels of awareness: for example, by helping them anticipate what they may see enroute, by pausing along the way, and by recall. By establishing what has sometimes been called an ''anticipatory set,'' the teacher prepares students in such a way they will find value in the process of reaching a destination (a not-always-appreciated part of a field trip).

Finally, camping provides another rich avenue for spatial experiencing. Although some school districts have provided camping experiences on a systematic basis (usually as part of the curriculum for specific grade levels,) and have maintained camp facilities on a permanent basis, these are not necessary prerequisites to providing such opportunities for students. When such experiences are not integrated into the curriculum, a camping experience without these resources calls on an additional commitment of energy and dedication, not only for the teacher concerned, but for the students' parents as well. Public camping areas are, of course, not universally available, but their presence is generally underutilized by individual classroom groups.

## IN SUM

The strange fact of the matter is that the role of geography in the school curriculum is at once anomalous and ubiquitous. Geography lacks a clear identity, partly because of the nature of geographic inquiry in its more formal sense; that is, there is no simple or even clear answer to the question, *What is geography?* When educationists seek what the many answers to this question might mean for teaching and learning, they are hampered not only by their own lack of knowledge about the scope and content of modern geography as it is practiced professionally, but by the peculiar organizational structure of the typical school curriculum.

Nonetheless, by its very nature, geography is integral to all human inquiry. It is difficult, or even impossible, to separate what is geographic from what is not. In this sense, then, geography is everywhere in the school curriculum. The major problem, both for geographers and geographic educators and for all curriculum planners and teachers, is to find ways to acknowledge and act on this reality. A long time ago, John Dewey observed that ''perhaps the greatest of all pedagogical fallacies is the notion that a person learns only the particular thing he is studying at the time.'' Acceptance of this fallacy has led us to the kind of curriculum organization we know today. Faced with both a professional and a public attitude that worries over the acquisition of specific forms of information and equates skill with knowledge, it becomes difficult to argue that any ''subject,'' including ''geography,'' can be adequately represented in a curriculum that does not include a specific reference to its existence as a separate entity. This is the dilemma of geography and of teaching that evokes an awareness of spatial interactions and, in fact, lies at the heart of geographic inquiry.

## REFERENCES


Byklum, Daryl. November/December 1994. Geography and music: Making the connection. *Journal of Geography*, 93.6, 274–278.

Conniff, Richard. June 1992. To excel at ''O,'' study the map and run like hell. *Smithsonian*, 23.3, pp. 44–55.

Daly, John L. July/August 1990. Focus on geography—Team themes and field experiences. *Journal of Geography*, 89.4, 153–155.

Dimit, Claudia. 1989. Student attitudes toward science and the affecting personal, home, and school environment variables. Unpublished doctoral dissertation, Claremont Graduate School, Claremont, CA.

Donaldson, Margaret. 1978. *Children's minds*. New York: Norton.

Eichinger, John. 1990. High-ability college students' perceptions of secondary school science. Unpublished doctoral dissertation, Claremont, Graduate School, Claremont, CA.

Enedy, Joseph D. January/February 1993. Geography and math: A technique for finding population centers. *Journal of Geography*, 92.1, 23–27.

Gaarder, Jostein. 1994. *Sophie's world: A novel about the history of philosophy*. Trans. by Paulette Moller. New York: Farrar, Straus, & Giroux.

Goodlad, John I. 1984. *A place called school: Prospects for the future*. New York: McGraw-Hill.

Gutierrez, Esta Diamon, & Sanchez, Yvette. July/August 1993. Hilltop geography for young children: Creating an outdoor learning laboratory. *Journal of Geography*, 92.4, 176–179.

Hardwick, Susan W. November/December 1990. Using city directories to teach geography. *Journal of Geography*, 89.6, 266–271.

Kjellstrom, Bjorn. 1976. *Be expert with map and compass: The orienteering handbook*. New York: Scribner's.

Kunstler, James Howard. 1993. *The geography of nowhere: The rise and decline of America's man-made landscape*. New York: Simon & Schuster.

McNeill, Carol, Ramsden, Jean, & Renfrew, Tom. 1987. *Teaching orienteering: A handbook for teachers, instructors, and coaches*. Perthshire, UK: Harveys.

Miller, E. Joan. March/April 1989. Mark Twain in the classroom: Should we invite him in? *Journal of Geography*, 88.2, 46–49.

Nabhan, Gary Paul, & Trimble, Stephen. 1994. *The geography of childhood: Why children need wild places*. Boston: Beacon Press.

National Geographic Research and Exploration. 1994. *Geography for life: National Geography Standards 1994*. Washington, DC: National Geographic Research and Exploration.

Oden, Pat. July/August 1992. Geography is everywhere in children's literature. *Journal of Geography*, 91.4, 151–158.

Pellegrino, Charles. 1991. *Unearthing Atlantis: An archaeological odyssey*. New York: Random House.

Piaget, Jean. 1985. *The equilibration of cognitive structures: The central problem of intellectual development*. Chicago: University of Chicago Press.

Porter, Eliot. 1962. *''In wildness is the preservation of the world'': Selections and photographs of Eliot Porter*. New York: Ballentine Books.

Salter, C. L., & Lloyd, J. W. 1976. *Landscape in literature*. Resource Paper No. 76–3. Commission on College Geography of the Association of American Geographers, Washington, DC.

Slater, Frances, ed. 1989. *Language and learning in the teaching of geography*. London: Routledge.

Sarason, Seymour. 1990. *The preditable failure of educational reform: Can we change course before it's too late?* San Francisco: Jossey-Bass.

Spencer, Peter L., & Brydegaard, Marguerite. 1952. *Building mathematical concepts in the elementary school*. New York: Henry Holt.

Zirschky, E. Dwight. July/August 1989. Traffic light geography: A fifth-grade community project. *Journal of Geography*, 88.4, 124–125.

# 6

# Evaluating Progress toward Geographic Literacy

**Only 25% of American Adults Get Passing Grades in Science Survey**

Less than half of American adults understand the earth orbits the sun yearly, according to a basic science survey. Despite flubbing such questions, there is enthusiasm for research, except in such fields as genetic engineering and nuclear power.

Only about 25% of American adults got passing grades in a survey by the National Science Foundation of what people know about basic science and economics. . . .

The worst showing came when those surveyed were asked to define scientific terms. Only about 9% knew what a molecule was, and only 21% could define DNA. . . .

The national survey of 2,006 adults found that 72% believe science research is worthwhile. Only 13% took the opposite view.

Among college graduates, 90% thought the benefits of research outweighed the risks, while only 48% of those who did not complete high school felt that way.

Washington, D.C., from the Associated Press, May 24, 1996

## Americans' Science IQ

Here is the quiz used to determine how much American adults know about basic science issues. Answers are at the end, along with the percentage in the survey who answered correctly.

1. The center of the Earth is very hot. (True or False)
2. The oxygen we breathe comes from plants. (T/F)

3. Electrons are smaller than atoms. (T/F)
4. The continenents have been moving their location for millions of years and will continue to move. (T/F)
5. Human beings, as we know them today, developed from earlier species of animals. (T/F)
6. The earliest human beings lived at the same time as the dinosaurs. (T/F)
7. Which travels faster: light or sound?
8. How long does it take for the Earth to go around the sun: one day, one month, or one year?
9. Tell me, in your own words, what is DNA?
10. Tell me, in your own words, what is a molecule?

1. T, 78%; 2. T, 44%; 3. T, 44%; 4, T, 79%; 5 T, 44%; 6 T, 48%; 7. Light, 75%; 8. One year, 47%; 9. DNA, or deoxyribonucleic acid, is a large molecule in the chromosomes that contains the genetic information for each cell, 21%; 10. A molecule is the smallest unit of a chemical compound capable of existing independently while retaining properties of the original substance, 9%.

## ASSESSING THE "GEOGRAPHIC CONTENT" OF STUDENTS' MINDS—A BRIEF HISTORY

How does one come to know with any specificity what one might term the geographic content of a student's mind? As we have observed in Chapter I, it was not all that long ago that this very complex problem was thought to be quite a simple matter. Students memorized a textbook which, conviently enough, consisted primarily of a series of questions, the answers to which were expected to be verbatim recordings from the text itself. But it appeared that, to rid themselves of the resulting memory overload, the "scholars" of those days found a way to cope: by forgetting virtually everything that had come before in order to concentrate on the lesson of the day. Horace Mann attempted to reach a deeper understanding of the residue of instruction by devising a series of question (also mentioned in the first chapter of this book) that he believed would plumb his scholar's minds apart from the daily lesson. Of course, we know that the result of those tests caused him to bemoan what he perceived to be a very sad state of affairs.

However, Mann's test stood alone for many years. Instruction in American schools continued the tradition of memorization and regurgitation until the dawning of the Progressive movement. As we have also seen, Louis Terman latched onto the work of Simon Binet in Paris, converting the latter's attempt to identify mentally retarded children so they might be helped with special instruction into the notion of an *intelligence quotient*, the idea that ability is spread out over a range in which each person's score is perceived to be immutable, persisting for a lifetime. The idea of tests that would reveal accomplishment within a particular subject matter frame that might, in turn, be compared within and between individuals and groups arose out of Terman's work and his era's growing belief that all aspects of human behavior might ultimately be quantified and, thereby, understood.

The first effort in this regard where geographic knowledge was concerned involved, interestingly enough, Mann's earlier test. Two research/practitioners of the 1920s opted to utilize Mann's 1845 survey of students in Boston in an effort to prove the efficacy of the new Progressive emphasis on reasoning and understanding (Caldwell & Courtis, 1923). Emulating Mann's earlier survey, they administered individual tests over a wide range of subject matters. Here, of course, our interest lies in questions claimed to be within the realm of geography. Realizing that some of the questions in the 1845 test were not suitable "for twentieth-century children" since the knowledge base and the curriculum itself had changed in the previous 75 years, they nevertheless incorporated much of the original test.

The results of this first comparative study of educational achievement, not surprisingly in the light of what we now know about problems in achievement testing, were mixed. Caldwell and Courtis intended to show that the educational experience of students under Progressive forms of instruction, in which reasoning was valued along with what might be called the nuts and bolts of geography—the "facts"—and in what we would today call a "content analysis" of individual items on the test, would, in fact be superior. They argued that a number of the items did demonstrate that questions for which students would need some inferential power appeared to produce better, and even superior, answers. Illustrating this assertion, the researchers noted that "in 1845 only 40% of the children knew whether the nearest route to India is by the way of the Cape of Good Hope or by the Red Sea, while in 1919 78% of the children answered correctly" (1925, p. 87). They did not, however, make any judgments regarding the appropriateness of the questions. The assumption that the response to any particular question was, in fact, relevatory of a student's knowledge—in this case, of geography—was never called into question.

The 1920s and early 1930s, that period usually designated as the time of the scientific movement in education, included a broad expansion of the idea that measurement would reveal the schools' success in cultural transmission. The field of educational statistics emerged to fill the need for measures by which such learning might, presumably, be judged. The instrument of choice would be the so-called standardized test, and in a very short time, American educators (enamored with the idea their product, i.e., the learning of various kinds of information, ideas, and so forth, might be elevated to the level of scientific truth) began clamoring for tests that would cover the entire curricular effort. Unfortunately, where geography was concerned, this period also turned out to be the time when the idea of geography as a school subject increasingly became folded into the larger rubric, *social studies*. As a consequence, most of the tests falling within this period—and there were many—if they included geographically framed questions at all, incorporated them only as a small subtest of the larger whole.

Following World War II, Progressivism was largely left behind. Instead there dawned a period of intense criticism of the previously much touted American school system, the fountain of public enlightenment, the cradle of democracy, and the path to a better economic and social order (see, e.g., Bestor, 1955; Flesch, 1955; Rickover, 1959). This period culminated in 1957 with the launching of *Sputnik*, the final proof of the decried educational failures foisted on the American public by John Dewey and a romantic legion that was, inaccurately, described as having bought his ideas hook, line, and sinker (see, especially, Ryan, 1995). With Dewey's denoument came an increasing faith in the power of the standardized test as an objective instrument that would tell the truth about the quality of instruction being provided by the public schools.

Until this time the so-called objective achievement test had played a minor role in evaluating the work of the school, and even then it was employed more in the spirit of Binet than of Terman, serving to tell teachers and researchers what children knew rather than what they did not. Now, however, it began to assume its present role as a measure of a person's presumed effectiveness. By now as well, the idea of a test focusing exclusively on geography had largely gone by the boards, and of those few that purported to be exclusively geographic in nature, none were perceived as much more than tests of location or of familiarity with famous persons or horrific events. Such tests failed to include interpretive questions, and they frequently included outdated material. One evaluator was cited in the 1983 *Mental Measurements Yearbook* (Buros, 1983), in critiquing the Hollingsworth-Sanders Geography Test for grades 5 to 7: "it is difficult to imagine a geography test for the intermediate grades that places no emphasis on geography skills or relationships. No maps or graphs are used. Question after question asks students to recall locational and definitional information. Moreover, there are problems with the several forms of the test as an adequate measure of geographic information" (p. 1309)

The failure in the 1960s to coalesce the various interest groups within the High School Geography Project stands as a symptom for all that went wrong in keeping a focus on the geographic strand of the school curriculum. Responding to the demise of a particular interest or concern in that regard, Buros et al. dropped *geography* as a separate category in their classification system in 1965, including it thereafter, sporadically, as an aspect of social studies.

## GEOGRAPHIC LITERACY AND THE CURRENT MOVEMENT TO REFORM THE SCHOOLS

The American public has become increasingly addicted to the idea of testing. To that end, the U.S. Congress, in 1969, authorized the U.S. Office of Education to institute a systematic program designed to conduct an ongoing, nationwide assessment of student achievement. Under its Office of Educational Research and Improvement, Congress contracted with the Educational Testing Service (ETS)—located in Princeton, New Jersey and a private corporation not affiliated with but often confused with Princeton University—to develop and administer the testing program. There has since periodically appeared a publication called *The Nation's Report Card*, through which the findings of these tests are disseminated. The National Assessment of Educational Progress (NAEP), the testing program developed by ETS that provides the data for these reports, has grown over the years so that it now covers representative populations in every region of the United States. It is not a test designed for widespread use. Rather, it is intended to provide a snapshot of what is thought to be a part of a larger whole. Administered every two years, over time they incorporate each of the traditional "subjects" of the curriculum—a strong deterrent to the idea that combining subject matters, as in the notions of *fusion*, *broad fields*, or other concepts of rethinking traditional curriculum boundaries (as in *social studies* itself), might be an efficacious model for learning.

*The Geography Learning of High-School Seniors* (ETS, 1990) reports a subsample of a more comprehensive test administered in 1988. This marked the first time ETS included a section on "Geography," and in that particular year, only seniors were included. The NAEP protocol presently calls for tests to be administered at three grade levels: 4, 8, and 12. "Geography," along with "History" and "Reading," comprised the 1994 emphasis. In prior years, the "subject areas assessed" by NAEP had not included "Geography" as a separate entity; rather, the rubrics were "Social Studies" or "Citizenship/Social Studies" (1971–72, 1975–76, 1981–82), "Literature and U.S. History" (1986), and "U.S. History" (1988, 1994). The designations are reflective of the trend toward the further separation of subjects at the elementary school level and a reaffirmation of the traditional secondary school curriculum, just the opposite of what is being seen in American college and university curriculums. Overall, we may be seeing the influence of the geographic community, and particularly of the National Geographic Society, in emphasizing geography, however we may think of it, in the school curriculum. Since the history of assessment in this area is so short, it is not possible to predict how often the NAEP will include it within its framework of tests.

## Establishing Baseline Data for Geographic Literacy

The 1994 assessment focus on geography resulted in the development and administration to a substantial number of students of a new test. The primary purpose of the test at this stage is to establish baseline data on which future comparisons about the effectiveness of the school curriculum may be validated. We can therefore expect it to take on an important presence in both assessment and curriculum building in future years. Still, ETS acknowledges:

> Geography, unlike many other subjects, is not taught as a separate topic in all schools. It can be learned in a wide variety of settings—not all of them restricted to the classroom. In developing the Geography Assessment Framework, great care was taken to ensure that the material broadly reflects what it is that students should know and be able to do in geography without specifying how or in what classes the material should be taught (U.S. Department of Education, n.d.)

Although it is conceded that ''geography'' may be learned in many venues, the school is ultimately being held responsible for teaching the concepts and abilities the test presumably assesses.

The mantra of the Educational Testing Service regarding each of its tests, including that of geography, is that they assess what ''schoolchildren know and can do'' (ETS, 1991). Within that frame, the test publishers assert that:

- *Fourth graders* should be able to use fundamental geographic knowledge and vocabulary; be able to illustrate ways people depend upon, adapt to, and modify the environment; and be able to demonstrate how an event in one location can have an effect upon another location.
- *Eighth graders* should be able to describe physical and cultural characteristics of places and explain how places change as a result of human activity; be able to explain and illustrate how the concept of regions can be used as a strategy for organizing and understanding the earth's surface; and use information from maps to describe the role regions play in influencing trade and migration patterns.
- *Twelfth graders* should be able to discuss economic, political, and social factors that define space on earth's surface; be able to relate the spatial distribution of population to economic and environmental factors; and, using maps and tools, be able to report both historical and contemporary events within a geographical framework. (U.S. Department of Education, n.d.)

These comments (and those reported later in this chapter regarding the more detailed criteria upon which the test's development was presumably based) indicate quite clearly that the geographic determinism of the nineteenth century is far from dead.

These broad statements of competency are given much greater detail in the description of what ETS terms ''geography achievement levels,'' which were developed from an analysis of the results derived from the first major administration of the test. Regardless, the matter of *test validity* begs some consideration: does the NAEP test in geography test what it purports to test? As with other such test instruments, ETS lays claim to the notion of *face validity*, namely, the actual construction of the test was based on consultations with experts in judging what geography is and what students should know about it. Readers of the literature published in connection with the test and the reporting of its results are given little detailed information about the process through which this came about. One is told that the National Assessment Governing Board established ''the Geography Consensus Project to create a framework for the assessment's development, within which two other sub-groups functioned: a Steering Committee and a Planning Committee'' (U.S. Department of Education, n.d.). These groups, ''through a year-long nationwide consensus process, established what is most important for students to know and to be able to do in geography'' (ETS, 1991). The membership of the project's steering committee consisted of representatives from ''education, business, makers of public policy, and the general public'' (ETS, 1991). The project's planning committee was comprised of ''leading geographic educators and practitioners and the general public'' which developed a ''geography framework that is relevant to the future of the nation. . . . The framework was created with one question foremost in mind: *What fundamental geographic knowledge, understanding, and applications should students have mastered in order to be informed and productive 21st century citizens?*'' Table 6.1 reports the ''geography achievement levels'' derived from the above process (U.S. Department of Education, 1995).

**Table 6.1**
**Geography Achievement Levels**

## Grade 4

*Basic*: Students should be able to use words or diagrams to define basic geographic vocabulary; identify personal behaviors and perspectives related to the environment and describe some environmental and cultural issues in their community; use visual and technological tools to access information; identify major geographic features on maps and globes; be able to read and draw simple maps, map keys and legends; demonstrate how people depend upon, use, and adapt to the environment; and give examples of the movement of people, goods, services, and ideas from one place to another. In addition to demonstrating an understanding of how individuals are alike and different, they should demonstrate a knowledge of the ways people depend on each other.

*Proficient*: Students should be able to use fundamental geographic knowledge and vocabulary to identify basic geographic patterns and processes; describe an environmental or cultural issue from more than one perspective; and read and interpret information from visual and technological tools such as photographs, maps and globes, aerial photography, and satellite images. They should be able to use number and letter grids to plot specific locations; understand relative location terms; and sketch simple maps and describe and/or draw landscapes they have observed or studied. Proficient students should be able to illustrate how people depend upon, adapt to, and modify the environment, describe and/or illustrate geographic aspects of a region using fundamental geographic vocabulary and give reasons for current human migration; discuss the impact a location has upon cultural similarities and differences; and be able to demonstrate how an event in one location can have an impact upon another location.

*Advanced*: Students should be able to use basic geographic knowledge and vocabulary to describe global patterns and processes; describe ways individuals can protect and enhance environmental quality; describe how modifications to the environment may have a variety of consequences; explain differing perspectives that apply to local environmen-

**Table 6.1 (continued)**

tal or cultural issues; and demonstrate an understanding of forces that result in migration, changing demographics, and boundary changes. They should be able to solve simple problems by applying information learned through working with visual and technological tools such as aerial and other photographs, maps and globes, atlases, news media, and computers. They should be able to construct models and sketch and label maps of their own state, the United States, and the world; use them to describe and compare differences, similarities, and patterns of change in landscapes; and be able to predict the impact a change in one location can have on another. They should be able to analyze the ways individuals and groups interact.

## Grade 8

*Basic*: Students should possess fundamental knowledge and vocabulary of concepts relating to patterns, relationships, distance, direction, scale, boundary, site, and situation; solve fundamental locational questions using latitude and longitude; interpret simple map scales; identify continents and their physical features, oceans, and various countries and cities; respond accurately to descriptive questions using information obtained by use of visual and technological tools such as geographic models and/or translate that information into words; explain differences between maps and globes; and find a wide range of information using an atlas or almanac. Students should be able to recognize and illustrate the relationships that exist between humans and their environments, and provide evidence showing how physical habitat can influence human activity. They should be able to define a region and identify its distinguishing characteristics. Finally, they should be able to demonstrate how the interaction that takes place between and among regions is related to the movement of people, goods, services, and ideas.

*Proficient*: Students should possess a fundamental geographic vocabulary; understand geography's analytical concepts; solve locational questions requiring integration of information from two or more sources, such as atlases or globes; compare information presented at different scales; identify a wide variety of physical and cultural features and describe regional patterns. Students should be able to respond accurately to interpretive questions using geography's visual and technological tools and translate that information into patterns; identify differences in map projections and select proper projections for various purposes; and develop a case study working with geography's analytical concepts. In addition, students should be able to describe the physical and cultural characteristics of places; explain how places change due to human activity; explain and illustrate how the concept of regions can be used as a strategy for organizing and understanding the earth's surface. Students should be able to analyze and interpret data bases and case studies as well as use information from maps to describe the role that regions play in influencing trade and migration patterns and cultural and political interaction.

*Advanced*: Students should have a command of extensive geographic knowledge, analytical concepts, and vocabulary; be able to analyze spatial phenomena using a variety of sources with information presented at a variety of scales and show relationships between them; and use case studies for spatial analysis and to develop maps and other graphics. Students should be able to identify patterns of climate, vegetation, and population across the earth's surface and interpret relationships between and among these patterns, and use one category of a map or aerial photograph to predict other features of

**Table 6.1 (continued)**

a place such as vegetation based on climate or population density based on topographic features. Students should also be able to relate the concept of region to specific places and explain how regions change over time due to a variety of factors. They should be able to profile a region of their own design using geographic concepts, tools, and skills.

## Grade 12

*Basic*: Students should possess a knowledge of concepts and terms commonly used in physical and human geography as well as skills enabling them to employ applicable units of measurement and scale when solving simple locational problems using maps and globes. They should be able to read maps; provide examples of plains, plateaus, hills, and mountains; and locate continents, major bodies of water, and selected countries and cities. They should be able to interpret geographic data and use visual and technological tools such as charts, tables, cartograms, and graphs; know the nature of and be able to identify several basic types of map projections; understand the basic physical structure of the planet; explain and apply concepts such as continental drift and plate tectonics; and describe geography's analytical concepts using case studies. Students should have a comprehensive understanding of spatial relationships including the ability to recognize patterns that exist across Earth in terms of phenomena, including climate regions, time zones, population distributions, availability of resources, vegetation zones, and transportation and communication networks. They should be able to develop data bases about specific places and provide a simple analysis about their importance.

*Proficient*: Students should have an extensive understanding and knowledge of the concepts and terminology of physical and human geography. They should be able to use geographic concepts to analyze spatial phenomena and to discuss economic, political and social factors that define and interpret space. They should be able to do this through the interpretation of maps and other visual and technological tools, through the analysis of case studies, the utilization of databases, and the selection of appropriate research materials. Students should be able to design their own maps based on descriptive data; describe the physical and cultural attributes of major world regions; relate the spatial distribution of population to economic and environmental factors; report both historical and contemporary events within a geographic framework using tools such as special purpose maps, and primary and secondary source materials.

*Advanced*: Students should possess a comprehensive understanding of geographic knowledge and concepts; apply this knowledge to case studies; formulate hypotheses and test geographic models that demonstrate complex relationships between physical and human phenomena; apply a wide range of map skills; develop maps using fundamental cartographic principles including translating narratives about places and events into graphic representations, and use other visual and technological tools to perform locational analysis and interpret spatial relationships. Students should also be able to undertake sophisticated analysis from aerial photographs or satellite imagery and other visuals. Advanced students should be able to develop criteria assessing issues relating to human spatial organization and environmental stability and, through research skills and the application of critical thinking strategies, identify alternative solutions. They should be able to compile data bases from disparate pieces of information and from these data develop generalizations and speculations about outcomes when data change.

The world, as we all know, is full of shoulds and oughts. But learning, like politics, is (besides being local) the art of the possible. I find it notable that the experts engaged in the construction of this test seem not to have considered, or perhaps were simply unaware of, what appears to be intellectually possible—and what is entailed in learning about spatial interactions. If we give any credence to what research tells us about the acquisition of spatial concepts, and even more generally about intellectual development across the age span as well as the nature of the learning process more broadly viewed, then certain caveats need to be established before we assume the validity of this very serious and broad-scale effort to evaluate geographic learning and, by inference, the geographic strand of the school curriculum. Test construction in this, as in virtually every effort to develop standardized measurements of intellectual competence, takes on a life of its own, regardless of critics who might dare raise the question of validity.

ETS is best known for its monopoly in college admissions testing, but in recent years, particularly through its contracts with the U.S. Department of Education, it has also become a dominant force in the field of achievement testing in America's elementary and secondary schools. The ETS story is the stuff of legends, and while one might question the thrust and content of its many and varied assessment tools, its analyses of the data are sophisticated and complete. The NAEP geography assessment (U.S. Department of Education, 1995) is a particularly interesting example of the manner in which ETS functions, and it is suggestive of the power and influence this organization is having, not just where geographic literacy is concerned but, by extrapolation, of its highly influential national presence in decision making where curriculum development is concerned.

## The Test and the Findings

The NAEP geography national sample was drawn from students enrolled at the three grade levels (U.S. Department of Education, 1995). There were approximately 5,500 students in the fourth grade, 6,900 in the eighth grade, and 6,200 in the twelfth grade samples. This was a "national probability sample," selected systematically so as to include representatives from each of the major geographic areas of the United States and accounting as well for racial and ethnic makeup, gender, and type of school (public, private, and others, such as Department of Defense and Indian Affairs schools). Although the sample seems small in comparison to the number of students in the United States, we are here treated to the statistician's expertise in selecting a population that, at least statistically, is representative of the much larger population and stands the test of reliability in the sense that the chances are high that, in a test-retest situation, the findings would be virtually the same 95 times out of 100.

According to the ETS, the geography assessment posed questions in a two-dimensional matrix. *Space and Place, Environment and Society*, and *Spatial*

*Dynamics and Connections* constitute one dimension; the other is made up of *Knowing, Understanding*, and *Applying*. The latter list is reminiscent of Benjamin Bloom's *Taxonomy of Educational Objectives*, a hierarchical, but abstract, attempt to define different levels of thinking that found only theoretical use (1954–1964). Following recent developments in test construction, and in an attempt to get away from the traditional overdependence on multiple-choice questions, short-answer essay and problem-solving questions were developed for each of the grade-level tests. The analysis of the responses to individual items were evaluated in terms of three so-called achievement levels: basic, proficient, and advanced. These levels are also intellectual constructs and as such have yet to be verified as useful distinctions. Nonetheless, the scoring range for each test item as well as the overall evaluation for each of the grade level tests utilized this framework.

The results of the first administration of this test follow patterns predictable from its earlier administration to high school seniors, in which the effects of instruction are not particularly evident. Overall, very few students scored at or above the advanced level—less than 5 percent. Nationwide, almost a third of all students at the three grade levels scored *below* the basic level. Slightly over two-thirds of the student scores were at or above the basic level, while approximately a quarter scored at or above the proficient level. Also predictably, public school students scored less well than those enrolled in private or special schools, results mitigated by the fact that about 90 percent of the sample was drawn from public school populations. Boys outscored girls overall, although not dramatically so. Race/ethnicity patterns present a disturbing picture: white and Asian students outscored, in descending order, Pacific Islander, American Indian (each representing between 1 and 2 percent of the total population, therefore probably making this an unreliable comparison), Hispanic, and black students, with a pronounced separation of black students from all others. Students from the southeastern region of the United States scored less well than those from the three other regions (northeast, central, and west). Perhaps most telling of all the results, geography achievement levels measured by this test followed closely parents' educational level: the more education the higher the score in virtually every instance, again iterating the likelihood that teaching, so far, plays a minor role in the achievement of geographic literacy.

The attempt to specify levels of competence (i.e., delineating levels of achievement within responses for individual test items and translating that into a general finding) is a novel approach to test construction and, of course, entirely arbitrary, as was the setting of those standards that identified these ''achievement'' levels. At the individual test-item level, this approach makes possible a three-tiered assessment of those items that go beyond the traditional multiple choice or fill-in-the-blank type of question. Many of the test items were constructed so that responses could be categorized as complete, essential, or partial—again, arbitrarily chosen designations. An example selected from the eighth grade test to illustrate this method of evaluation reads as follows:

After we anchored our ships in the ocean and went ashore to explore, we marched west. The forest was so thick we could only travel three miles in the first two days. Then we came to the mountains and climbed to the top. A rushing river flowed west out of the mountains. Two miles further we came to the coast. It was obvious that the area we were exploring was an isthmus.

In the box below, draw a map of the region described above. Be sure to include all of the geographical elements mentioned in the description. Include a scale to indicate distances (U.S. Department of Education, 1995, p. 48)

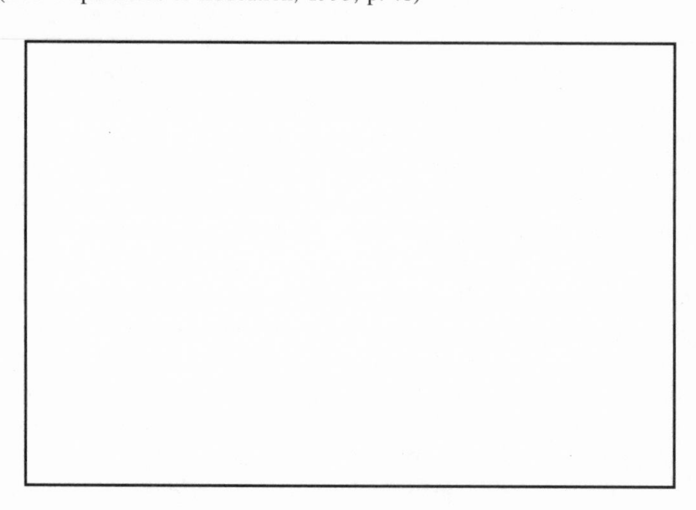

The protocol for scoring this particular item calls for evaluating an answer on the three levels:

A *complete* response includes an accurate map in which at least four elements are correctly placed.

An *essential* response includes a map in which three elements are correctly placed. The response may be a peninsula or an island.

A *partial* response includes a map in which at least two elements are correctly placed.

The remaining items on the test incorporate a combination of multiple-choice questions (primarily four-item choices rather than the preferable five items) and short-answer responses that can be evaluated as meeting one of the competency levels indicated above.

## TRENDS IN ASSESSMENT PRACTICES AND THEIR IMPLICATIONS FOR GEOGRAPHIC LITERACY

The NAEP geography test incorporates recent trends in standardized test construction: movement away from true-false and multiple-choice questions toward items that can be evaluated for meeting predetermined levels of completeness. The latter type of questioning presumably requires higher thought processes by encouraging reasoning over the recall of specific bits of information. While this trend in test construction appears to be a positive one, it is not without its problems, not least of which is that of scoring the responses themselves. The typical standardized test can be scored by anyone, and in fact, machine scoring has made possible the processing of large numbers of score sheets—the reporting of answers on the tests themselves in favor of individually completed answer sheets disappeared long ago. This has made it possible to process scores, even hundreds of answer sheets, in a very short period of time with minimum human attention, a development that has encouraged the proliferation of testing to increasingly large populations of students, which has, in its turn, made possible the development of statewide achievement testing of every student. With that has come the concomitant publication in the newspapers of test scores based on group averages and the inevitable comparisons between and within school districts. The consequence has been almost totally negative except, perhaps, for those few affluent enclaves where test scores always soar, a fact that raises questions regarding the relationship of testing to social, ethnic, and economic factors in the educational enterprise.

Tests that require evaluations of responses to individual items complicate the situation. Since answers are not simply right or wrong and because there are presumed to be levels of competence present in each response, the judgment of the person evaluating the test enters the picture. In the case of the NAEP geography test, it is very clear that a considerable amount of training would be required of anyone deemed qualified to judge the quality of the students' answers to any particular question. This goes beyond a simple matter of training for this particular test; the evaluator must be geographically literate as well. These conditions militate against the possibility that the test can be administered at any one time for a large population, and in fact, the so-called NAEP "baseline" testing utilized a small sample, even granting the care that went into its selection.

Another matter becomes important as the accompanying news story makes evident.

**Governor Signs Bill to Replace Controversial CLAS Test**

Gov. Pete Wilson on Monday bucked critics in his own party and signed a bill to set up a statewide academic testing program to replace the controversial CLAS [California Learning Assessment System] test that some parents had blasted as invasive of their privacy.

Considered by boosters to be the most significant education-related measure to emerge from the deeply divided Legislature this year, the bill . . . is designed to restore the state's ability to judge the performance of California's $27-billion school system while letting parents know how their children are doing.

"What I want to know is how the system is working," said Sen. Leroy Greene (D-Carmichael), the principal author of the bill. "If we are putting money into math or reading or whatever the thing might be . . . I want to know if it bought something."

State Superintendent of Public Instruction Delaine Eastin praised Wilson for signing the bill[:] . . . "Quality control is understood very well by business, but sadly, some in government do not recognize its importance. This assessment system is the key to quality control in public education in California."

In writing the bill, Greene said he sought to address problems that led to last year's demise of CLAS. . . . Many educators had considered CLAS a cutting-edge tool for assessing students' skills in solving problems, analyzing literature and writing, but some parents charged that it was inaccurate, that it ignored basic skills and that its open-ended questions invaded students' privacy. (*Los Angeles Times*, October 17, 1995)

Like the NAEP geography test, the so-called California Learning Assessment System (CLAS) sought to improve on the traditional right/wrong achievement test by introducing a multiple-value scoring system. However, it ran into another kind of value system, that of parents and a general public that guards what it perceives to be its "privacy" with what can only be described as a vengeance. The reaction of parents to qualitative assessment practices, which both the NAEP and CLAS tests represent, illustrates the often-made point that the educative process has the potential for being very dangerous (i.e., dangerous to those who fear the process of valuing in someone other than themselves, especially when the "other" is a young person). The language we choose to illustrate this point is all too obvious: we speak of an "invasion" of privacy, as if the problem were equivalent to war itself. Thus, although the present NAEP geography test may appear to be more benign than, say, one that more clearly, perhaps, involves evaluating the quality of one's writing or the interpretation of a piece of writing (the selection of which may, on its face, be controversial to some minds), in a more developed form it may not be as value-neutral as one might think.

Another danger becomes evident. Although recent trends in test development are widely believed among professionals in the field of educational assessment to be a forward step in remediating the ills of the right/wrong tests as we have known them, it is clear that there is a substantial public that harbors less-than-benign opinions about probes that are more successful in getting at quality over quantity in student thinking. It is consequently likely that conservative pressures will inhibit the expansion of this innovation in assessing student progress. If this is true, we may also expect intensive testing on a statewide basis, as is now the case in many states, to continue, and perhaps to intensify even further. Attempts at evaluation practices that seek to enhance qualitative assessment practices are time-consuming and, as a consequence, not amenable to mass testing, while theoretically more sensitive tools for evaluating the effectiveness of teaching efforts may very likely fall by the wayside or at least find it hard going as a substitute for the conventional practice of frequent testing of large numbers of students.

Lost in the fervor of mass standardized testing is the fact that every test is nothing more than a sample of all the things students ought to be learning during their days in school. However, every teacher and school administrator understands that first things must come first, namely there must be satisfactory test scores or there will be an up-in-arms public demand that the school return to "the basics," namely, those things that will make possible the achievement of the desired results: the test scores themselves. To achieve this goal, there will be pressure to teach to the test, that is, if not to the specific test, at least by giving increased attention to the areas in which the tests are directed. Since achievement testing generally deals with the "skills curriculum," any hope that there will be balance in the curriculum is likely to be lost, particularly where test scores are perceived to be unacceptably low. In the case of the geographic strand of the social studies curriculum, given the history of assessment in both its broad and narrow sense (i.e., *geography*, in its more traditional sense, versus social studies as an area encompassing concepts drawn from geography), it is not surprising that there is no sustained record of evaluating student learning in this aspect of the curriculum. Any informed observer of the history of the social studies will attest to the fact that, as an area of emphasis, its presence within the typical school curriculum, both at the elemenary and secondary levels, has declined as the intensity of achievement testing in the so-called "fundamentals" has increased.

Yet another matter deserving attention deals with the amount of time achievement testing takes away from learning. It is generally conceded that, since testing seeks to determine (within the limits suggested) how much has already been learned, the time devoted to discovering what this might be is not in and of itself much of a learning experience. Some have likened it to that of the gardener who systematically pulls up the plants to find out how they are growing and then puts them back in the ground again to, perhaps, continue to survive and grow, but with the effects of their uprooting always evident. Certainly, as achievement testing has assumed its nonpareil position on the educational scene, students spend many hours—all told, many days—providing evidence of what they have or have not learned during the previous months or even years. Few parents (or the public generally) appreciate this very large drain on the educative time of students, nor do they understand that the situation is likely to worsen. It is perhaps yet another strange phenomenon surrounding attempts at school improvement that the CLAS test, which was designed, as with virtually all recent attempts at test improvement, to make "testing" a natural part of the daily rhythm of learning, should rather be viewed as a transgression on student (and, perhaps more tellingly, on parent) privacy and be cast aside as a consequence.

If by some miracle these pressures were to mitigate, it is conceivable that tests such as the NAEP geography test would, if used selectively and if the test results were viewed as suggestive rather than as criteria for teaching, find a valued place in the policy-making side of curriculum development activity. Lost in the fervor of testing every student at specific intervals throughout a school career is, after all, the fact that achievement tests are not diagnostic regarding individual behavior. Their results, even at their best, are only suggestive for what one might call the larger picture: where school professionals create maps suggesting what might constitute a balanced curriculum for their particular clientele.

## REFERENCES

Bestor, Arthur. 1955. *The restoration of learning: A program for redeeming the unfulfilled promise of American education.* New York: Knopf.

Bloom, Benjamin. 1954–1964. *Taxonomy of educational objectives: The classification of educational goals.* New York: David McKay Company.

Buros, Oscar K., ed. 1983. *The mental measurements yearbook.* Highland Park, NJ: Gryphon Press.

Caldwell, Otis W., & Courtis, Stuart A. 1925. *Then and now in education 1845–1923: A message of encouragement from the past to the present.* Yonkers-on-Hudson, NY: World Book Company.

Educational Testing Service (ETS). 1990. *The geography learning of high-school seniors.* Report No. 19-G-01. Princeton, NJ: Educational Testing Service.

Educational Testing Service (ETS). 1991. *The NAEP guide: A description of the content and methods of the 1990 and 1992 assessments.* Prepared by Educational Testing Service for the National Center for Educational Statistics. Washington, DC: U.S. Department of Education.

Flesch, Rudolph. 1955. *Why Johnny can't read—And what you can do about it.* New York: Harper.

Rickover, Hyman G. 1959. *Education and freedom.* New York: E. P. Dutton & Co.

Ryan, Alan. 1995. *John Dewey and the high tide of American liberalism.* New York: W. W. Norton.

Trace, Arthur S. 1961. *What Ivan knows and Johnny doesn't.* New York: Random House.

U.S. Department of Education. National Academy of Education. 1987. *The nation's report card: Improving the assessment of student achievement.* Report of the study group: Lamar Alexander, chairman; H. Thomas James, vice-chairman and study director. Washington, DC: U.S. Department of Education.

U.S. Department of Education. National Assessment Governing Board. (n.d.). *Preparing citizens for the 21st century: Geography: Learning about our world.* The 1994 National Assessment of Educational Progress in Geography. Washington, DC: U.S. Department of Education, National Assessment Governing Board.

U.S. Department of Education. Office of Educational Research and Improvement. October, 1995. *NAEP 1994 geography: A first look: Findings from the National Assessment of Education Progress.* Washington, DC: U.S. Department of Education.

# 7

# Geographic Literacy—Contexts for the Future

In the preceeding pages I have emphasized a central theme that is essential to thinking about a curriculum disposed to raising literacy levels where geographic understandings are concerned. We started with an often-overlooked fact and derived from it some implications for teaching, which are on the surface rather obvious but are also critical to changing the practice of teaching where geographic concepts are concerned. The overlooked fact has to do with the very nature of geography as a field of inquiry. Geographers are always eager to point out that their discipline is characterized, not so much by a specific body of subject matter or information as it is by a way of thinking about phenomena within a spatial context. Geography shares this uniqueness about the importance of method over subject matter in defining its nature with history, which also is open to the study of virtually any topic or question. In the case of history, interactions and interrelationships over time provide the framework for analysis. Historians report their observations in almost every instance through some form of verbal expression. In addition to that avenue for reporting data, the geographer relies on the map, with its specialized symbols and its reliance on mathematical representation, as a way of organizing and then finding significance in the information represented thereon. Although virtually anyone may write history or report on geographical phenomena, specific professions have come into existence that systematize, and presumably, although hardly always, elevate the quality of their work over that of the amateur practitioner of these "disciplines."

I have stated my belief that a major mistake in our attempts to teach geographic concepts has been to distinguish between what geographers seemingly do, namely, seek to understand interactions within a spatial environment, and what we think it is students ought to learn. We have separated "real" geography (or geography as professional geographers practice their profession) from instruction designed for the uninitiated, which we have generally called *school geography*. As a result, geographic information has become disembodied from its context of spatial inquiry, effectively disconnecting it from the utility required if it is to become a working part of an individual's intellect.

It would be naive to think that geography as a professional activity can be translated directly into the classroom, either in the elementary or secondary school, and perhaps not for many in our college classrooms. Mature geographic inquiry we leave to those people who practice the profession of geography. However, that does not mean that the process of geography is not available to even the youngest minds, *provided* we pay attention to their interests and their intellectual capacities as far as we know them to exist. What we ultimately should be looking for, then, is what one might think of as a habit of mind where geographical ideas are concerned, namely, the acquisition of a sense of space within which it is quite natural to organize data; this is the essence of thinking geographically.

It is all too clear how notoriously unsuccessful we have been in this endeavor. Adults, who represent the end product of the American school system (be it in public or private hands), simply do not possess the tools to think about issues, problems, or questions that are of a spatial nature, thus providing vivid evidence of the failure of the schools to accomplish this portion of their overall task. Yet the world is full of such problems. For example, much of city government revolves around problems regarding its uses of space. It follows that an informed electorate is essential in reaching equitable decisions regarding its uses, both public and private. A geographically illiterate community is of little help, and is often a hindrance, to the development of wise public policy regarding questions of land use. Moreover, this is only one of a myriad of issues, problems, or opportunities, that a spatial sense ennobles.

The geography curriculum is not alone in its failure to achieve the intended goal, of course. Our accomplishments in the teaching of history are perhaps even more abysmal, given the much greater amount of time over their school career that students are required to "study" historically significant events and circumstances. This failure serves as a useful parallel since it is also one in which "facts" take precedence over any effort to couch this equally important field of inquiry in terms of process, what historians called the *historical method*. If we are indeed to avoid the mistakes or errors of the past, which history is said to accomplish, we can all acknowledge the importance of being historically minded. Although history can justify itself with this very pertinent and often-heard paean, for which geography unfortunately has no close parallel, it nonetheless shares a similar responsibility. Urban sprawl, decaying center city cores, environmental issues of all kinds, and international affairs are played out in a spatial environment that demands but seldom secures literate attention.

Why is it, then, that instruction concentrates so intensely on the Mozambique Question: the "where is it, what is it?" kind of instruction—and on the kind of tests designed to find out whether students have solved this riddle which is so *product* oriented? A good part of it lies, I believe, in the persistence of an approach to instruction that emphasizes a quantitative definition of knowledge over a qualitative one. Schools are essentially conservative institutions, and therein we find another piece of the puzzle—one that explains, at least in part, why a "product" curriculum is difficult to dislodge. Of all the subjects taught in our schools, "the social studies" (perhaps arguably when one notes the resistance to innovation in the teaching of mathematics) are the most controversial for they are, by definition, concerned with values as well as with subject matter per se. Those who have the highest stake in the schools—teachers/administrators and parents, not to mention the general public in the long run—on the one hand want a qualitative education, but on the other fear a curriculum that might raise questions that appear to conflict with values held by the adult members of the community. These values are shaped by remembrances of things past, which adult minds now hold in a romantic light because parents (along with those who take on the problems of education for various personal reasons) having escaped the toils and troubles of compulsory school attendance, are generally in possession of a more benign view of their experiences as students. Meanwhile, teachers and their administrators often also join in misinterpreting that past, but with a special sense of knowing that their association as professional people bestows on them, which protects the status quo because the prospect of change is in itself rather frightening. It is in this context, then, that we face the popularly shared notion that all would be right with the geographic world if only a proper place were to be made in the curriculum for instruction that could be labeled specifically as *geography*, more or less without regard for how that teaching proceeds. In other words, we have not given up on the notion that what is taught is, necessarily, being learned, and that therefore, if time is set aside for this specific subject, then everything will indeed be fine.

## THE LARGER SCENE

Thus we arrive at the widely held view that, with the pronouncement of the existence of something called *the social studies*, geography lost its proper place in the school curriculum. Without that niche, it is believed, geography was doomed to exclusion when, in 1916, The Social Studies Committee (Saxe, p. 1) proposed the term *social studies* as more descriptive of what a school curriculum should include where matters involving the social education of students were concerned. Believing this as fact places entirely too much authority on the power of language; simply proclaiming something does not make it so. And while professional policy makers might have embraced the idea of social studies—they were really more concerned about the kinds of skills and information a socially responsible citizenry required than in defining some new kind of curriculum animal when they hit upon this term—classroom practice only very gradually shifted from the traditional, single-subject curriculum to one that dared to break down this form of classroom organization. Of course, that kind of fusion or integration did not go very far. Although schooling at the turning of the twenty-first century is in many ways very different from that of 100 years ago, even in today's typical elementary school classroom, where there is more often than not a single teacher for a group or class of children (unlike the secondary school where different subjects are taught by different and presumably specialized teachers), the school day is invariably organized around a series of separate "subjects," just as is universally done in the high school. (Larry Cuban makes this point particularly well in his 1993 book, *How Teachers Taught: Constancy and Change in American Classrooms, 1890–1990*).

Saying that, it is still true that geography never was able to gain equal footing with history. And it did not fare very well in the overall curriculum scheme as the years went by. Initially the problem lay in the fact that the geographic profession was itself in some disarray; thus, geography as a curriculum concern in the schools did not have the advocates enjoyed by history (through the American Historical Association) or by proponents of a strong civics curriculum (through the American Political Science Association). The comparable academic organization, the American Association of Geographers, was not formed until 1904, too late to participate effectively in the attempts at curriculum reform that occupied the decades at the turn of the last century (described in Chapter 1). A second factor tending to weaken the prospects of an expanding interest in the teaching of geography lay in the fact that the field of geography suffered from the destabilizing effects of the trend toward specialization generated by the knowledge explosion that began with Darwin's publication of the results of his voyage on the *Beagle*, a phenomenon that appears to be compounding itself geometrically (certainly with no end in sight nor even a sign of leveling off). As field after field split off from its home base—the earlier and most important division having occurred with geology finding a life of its own outside the previously all-encompassing walls of geography—we witnessed the launching in rapid order of such specialized areas as meteorology, climatology, seismology, oceanography, hydrology, and countless others—fields that even now are birthing their own highly specialized subfields. In one sense, then, geography has suffered an implosion; while finding its overall dimensions subject to reduction, what remains—the realm of spatial analysis—has been enhanced by the same knowledge explosion affecting every aspect of human knowing.

With relatively few practitioners of geography at an academic level, the problem of communicating with the teaching profession, either by influencing curriculum decision making or, more directly, within the bounds of the rapidly expanding practice of actually educating teachers for their work in classrooms, would therefore be part of the problem from the start. However, it would be compounded with the adoption of Paul Hanna's (Hanna, 1987) concept of the Expanded Communities curriculum, first delineated in the Virginia Study, (Virginia State Board of Education, 1934) and institutionalized in the years following with the publication of textbooks based on it. The widespread adoption of the concept of *expanding communities*, with its emphasis on the *basic human activities* rather than on the idea of communities acting within an ever-broadening spatial environment, won the day. Oddly enough, and illustrating the continuing faith in the power of history in building citizenship, Hanna's schema bowed to the power of the history lobby by inserting a series of interruptions at the fifth, eighth, and eleventh grade levels (state history was often interjected at the fourth grade, was well), during which students would study American history, along with a less rigorous attention to European history and its Judeo-Christian antecedents.

It is instructive in this regard to document history's record of failure since it furthers my argument that teaching that focuses on the near term—on the product rather than the process—will likely fail, no matter how much time we devote to its teaching. For example, the NAEP test in history (the equivalent to the NAEP geography test), revealed that at age seventeen, when some 80 percent of those taking the test were enrolled in the second semester of their high school American history course, nearly two-thirds of the sample did not know that the Civil War occurred between 1850 and 1900; almost 40 percent did not know that the *Brown* decision held school segregation unconstitutional, and another 40 percent did not know that the East Coast of the United States was explored and settled mainly by England and that the Southwest was explored and settled mainly by Spain. Another 70 percent did not know that the purpose of Jim Crow laws was to enforce racial segregation, while 30 percent could not find Great Britain on a map of Europe (Ravitch, 1989, p. 52). This is not a singular horror story. A test similar to the new NAEP geography test, which was designed around an effort to institute qualitative evaluative procedures, has not yet been fielded for history and very likely will not be, given the public (and political) predilection toward asking questions that entail quantitative measures of achievement. If such were to be become a reality, the state of historical literacy might not appear in such stark know-or-not-know terms as it now appears; however, there would also not likely be any reason for celebration, any more than is now the case where true geographic knowledge is concerned.

I have mentioned that *social studies* in whatever guise, as a single subject or an area incorporating information from several fields, is often perceived as a dangerous ''subject,'' a point that needs emphasis if we are to achieve a reasonably full understanding of the problems involved in achieving curriculum reform. The minute one steps beyond the safe confines of those specific bits and pieces of information that make up most teaching in this area, trouble arises. For example, when assessment devices are used that seek to get a reading on the quality of student thinking, concerns begin to be expressed about invasions of privacy. When teachers enter this arena, as many have to their regret, sparks often fly. And virtually any aspect of the curriculum may be subject to attack. Not a few books have been banned, sometimes from school libraries, but in other instances, topics deemed too sensitive have been excised or deleted ahead of time because of their ''controversial'' nature. How might a community react, for example, if the subject for study happened to be the geography of AIDs, a significant topic, which can be studied profitably within a spatial context? Past experience does not provide a hopeful guide. I have already made note of attempts to subvert the study of social problems and issues. For example, in the 1930s, textual material dealing with the consequences of the Depression—the existence of breadlines, the strikebreakers, the consequences of farming practices that resulted in the Dust Bowl and the migrations out of the Midwest—were censored as subject matter taught within the four walls of American classrooms. Much later, the political uproar that accompanied an experimental curriculum brought political repurcussions extending all the way to the Congress. We have not put curriculum problems like this behind us by any means. There are, unhappily, many other examples.

Young children are not small adults, however we may contradict that fact in everyday life, and clearly this must be taken into account when planning for teaching. Yet the irony in this fact remains, as several recent books dealing with what is being called "the disappearance of childhood" attest. They provide a chilling commentary regarding the present trend toward reversing a long tradition, which began with Jean-Jacques Rousseau, Johann Pestalozzi, Johann Herbart, and Friedrich Froebel in the 1700s, and continued with Maria Montessori and the kindergarten movement of the nineteenth century (Postman, 1982; Suransky, 1982), one very evident manifestation of which is found in the trend toward formalizing instruction in the kindergarten and, even more telling perhaps, the growth of "academic preschools" in the private sector. Testimony to the growing tendency of treating children as miniature adults who should be treated accordingly is found perhaps most dramatically, however, in policies that increasingly are authorizing the prosecution of teenagers as adults. This is occurring in the face of the near-unanimous opinion of child development experts who have long counseled against abandoning established ways of dealing with young offenders, since juvenile problems are more often than not direct reflections of parental and environmental effects, circumstances that are beyond the control of the child and that, theoretically at least, are subject to positive modification. We see the same scenario being played out in what we might think to be more benign environments—school classrooms rather than courtrooms—where pressures for the setting of standards that all must achieve are being demanded. The hope expressed in support of such proclamations is that they will somehow lift overall performance levels, the criterion for which remain rooted in the proposing of detailed statements of what students "should know and can do." Doing, in this sense remains a far cry from what John Dewey meant when he spoke of *learning by doing*. Nonetheless, this is proposed as the standard for judgment, not only for geography, but in other areas as well. It is assumed, along the way, that standard tests will be developed to determine how well the so-called standards have been met. Given the difficulties inherent in mass testing, there is little doubt that the tests that are developed will as a consequence necessarily continue to contain much of the traditional reliance that such measures have always included, in which emphasis on the quantitative plays an important role, albeit not quite as exclusive as in the past.

We have Jean Piaget to thank for putting some scientific flesh on the bones of these long-held assertions regarding the uniqueness of childhood where learning is concerned. Although most who know of his work think of him as a psychologist, Piaget preferred to be known as an epistemologist, since he perceived himself primarily as a classifier of the dimensions of human intelligence rather than one who explained why people thought as they did. It was Piaget who would detail that development, validating it by posing problems for children that would demonstrate qualitative changes in children's thought over time. I have provided considerable detail about Piaget's work in discovering this sequence where spatial understandings are concerned (in Chapter 3) because it is without question seminal in helping us appreciate some of the impediments and opportunities involved in the acquisition of spatial understanding. In the years immediately following the publication of Piaget's important work, a number of researchers replicated and/or expanded on Piaget's research. Unfortunately, little has been done in this regard over the past decade or so, a development which suggests an eerie if not factual conjoining of the trend toward blurring childhood with adulthood.

A major criticism of Piaget's research is to be found in Margaret Donaldson's book, *Children's Minds* (1978). Donaldson argues that one gets different ideas about immature thinking if the problems and questions put to children are phrased so as to better take into account the child's worldview. She criticizes Piaget's experimental designs, as have others, as being too artificial, saying that in reality, children can solve spatial problems such as those posited by Piaget (e.g., the mountains experiment) when the questioning and situations are altered in favor of children's own reality. Piaget, the criticism goes, was too rigid in designing his experiments and failed to make them sufficiently child-like; perhaps one can see some reasoning here in Piaget's eschewing of the label of psychologist for that of the epistemologist. Lamentably, despite the apparent validity in Donaldson's criticism, her research on neo-Piagetian theory has also not been expanded upon to any significant extent either in our research institions or in the field.

Like all significant discoveries, initial theories tend later to be viewed as needing expansion and revision. Moreover, as is usually the case, fundamental contributions to understanding, despite the inevitable revisionist activity, retain their integrity. This is, I believe, the case where Piaget's work is concerned; his basic discoveries regarding the existence of qualitative changes in reasoning survive in their basic outline. The major problem with Piaget's construct, or schema, regarding the course of mental development is that they reflect the determinism that characterized much of the thinking in geography and in intellectual life generally before and during the earlier years of the twentieth century—a failure in reasoning from which we have yet to extricate ourselves. Piaget came to maturity during these years, so it is not surprising that his worldview might be painted with this brush. Thus we see the genesis of a conceptual framework that envisioned a march toward intellectual maturity as one progressing along a number of fronts—the ability to conceptualize spatial relationships among them—which, although presenting a kind of "broken front" developmental pattern unique to each child, was inexorably developmental and nonreversable. Once a stage was achieved, there was no going back. Piaget rejected the idea that the particular circumstance might influence whether a child might be able to reason within one or another of the assigned stages.

Most of Piaget's research centered on young children through the ages of 9 or 10. From these earlier studies he extrapolated much of the characteristic of the Concrete Operational Stage, going even further out on the limb in describing "hypothetical" reasoning as the final and most advanced stage in intellectual development, available only to the adolescent and young adult (beginning around the age of 12). Although subsequent research has tended to confirm the existence of these stages, it remains largely understudied. What evidence we do have surprises us by suggesting that it occupies a much longer period of development than Piaget had theorized. Also, the possibility that a significant number of secondary school and college-age students may be more accurately described as remaining concrete operational in their thinking than might otherwise be presumed lends a cautionary note to teachers of those age levels: there is a reasonable prospect that instruction at these levels is assuming powers of abstract reasoning out of proportion to the number of students who may be able to engage in it (Downs & Liben, 1991, p. 304).

Indeed, early on, when Piaget's ideas were first making an impression in the United States—they experienced considerable difficulty breaking into the mainstream of American psychological thought, conflicting as they did then, and now, with the widespread acceptance of behavioral theory based on the stimulus-response paradigm—some researchers engaged in a number of cross-cultural studies raised the possibility that some cultures, notably those in the lesser developed world, were structured in such a way that few, if any, of their people demonstrated much development beyond the Concrete Operational stage. The causal factor here, some have suggested, is that the kind of experiences needed to stimulate the development of theoretical or hypothetical thinking are not as widely available as they are in developed countries. The implications of these admittedly underinvestigated possibilities within a society with quite clearly demarked social strata or where large numbers of immigrant populations from underdeveloped countries are found have not been explored. The situation in American society may not be exempt from this speculation.

Piaget and others stimulated by his research who have sought to understanding the qualitative aspects of thinking in the context of growth and development have provided information deserving of consideration by teachers and geographers interested in geographic education. Although Piaget's rigidities should be acknowledged—we are each a product of our time, whether we acknowledge it or not—there is no question that his conceptions bear those marks of originality and validity that characterize every major intellectual accomplishment.

## THE "CYBER" WORLD AND GEOGRAPHIC LITERACY

Although it may be argued that humankind, at least in the Western world, entered the information age'' with the invention of movable type, that term has now become synonymous in the popular mind with the unimaginable changes wrought by revolutionary advancements in the application of electronics in communication during the very short period of two or three decades. Just as there yet seems to be no limit to this technological revolution, so do people of all stripes seem tempted to hold the greatest of expectations for its power not simply to reshape society but to solve all its problems as well. Politicians of every stripe are being lured into the electronic tent, where education, particularly, is seen to be the most likely to benefit. Policy makers are praising the potential power of the computer, the Internet, the CD-ROM (television, the typewriter, and projecting devices of various kinds, among other technological marvels, having sung their own siren song in previous years), and now every classroom will be hooked up to the rapidly expanding "cyber" world by the turn of the century. With over a million classrooms in the United States, the goal now so glibly stated is nothing less than daunting should things turn out the way proponents of technology-as-answer predict.

What the future holds only the most prescient might imagine, but there is room for some doubt, and even pessimism. What has in its earlier developmental stages seemed full of promise is now giving rise to critics who do not see the rosy glow that others claim for cyberspace. One of the most articulate is Neil Postman. In his book, *Technopoly: The Surrender of Culture to Technology*, he argues that we are being threatened by an information glut in which the tendency is to value "information" over reasoning—a postmodern development likely to further reinforce the concept of a quantitative over a qualitative curriculum (1992). *Technopoly* is the term Postman invented to describe how this massive new ability to collect, store, and distribute information has resulted "in the deification of technology, [meaning] that the culture seeks its authorization in technology, finds its satisfactions in technology, and takes its orders from technology" (p. 71).

Geographers themselves are feeling the consequences of *technopoly* in ways previously unimagined. I refer here to the development of what some are calling a new science, indeed one that may replace geography as we have known it and a development with profound implications for geographic literacy. I refer here to *geographic information science* (GIS). Some refer to it as a "system," since it is a development within cyberspace devoted to the "capture, storage, analysis, and visualization of geographic information" (Sui, 1995, p. 580). It is, in any event, a data bank composed of all sorts of information—from satellite imagery to more mundane statistical data—to which anyone may turn, given the skills necessary to unlock it from the Internet. Because it takes advanced computer skills to release GIS information from its technological hiding place, colleges and universities are increasingly offering courses devoted to this skill alone—in this sense, it becomes a ''science'' just as another recent newcomer to the college curriculum, *computer* or *information science*, where study of the technology of cyberspace is viewed as valuable in its own right and has become an accepted new ''academic discipline.'' The concern within the geographic community is that knowledge of the technicalities in operating with the GIS data bank will obviate any interest in dealing with the traditional concerns of spatial analysis, which is still generally considered the raison d'être of geographic inquiry within the profession. Instead, they see a huge body of data being ''accessed,'' as the retrieval process is sometimes referred to, primarily to solve highly specific (so far, mostly economic) problems. There seems to be little question that geographers will need to become skilled in the techniques of GIS; the more important issue is whether, after all is said and done, there will be people who utilize that data to study geographic (spatial) phenomena in its broader context—whether there will be *geographers* in the sense we have known of this profession over the recent past. The field of geographic inquiry has served as a ''mother'' to other sciences. Thus, the fear where GIS is concerned is that a field already fragmented by defections brought about by specialization may now suffer the ultimate blow to its integrity should the emphasis shift too far toward gathering and dispensing data as an end in itself—whether it will be viewed primarily as a source of data/information than an enriched source for analyzing and interpreting geographic phenomena (Sui, 1995, p. 581).

The problem for geographic literacy in this situation lies with the fact that the geographic profession provides the reservoir of information and skill needed in teaching and learning about spatial phenomena. Although the geographic profession has not, over the years, been able to be as helpful to the teaching profession as one would like, there is real danger that, should the discipline be further fragmented, what little it is able to do will be much less than is presently the case. One teaches what one knows. If students are to be geographically literate, their teachers must be as well. If the materials teachers have at hand for teaching are to be improved, those who have a hand in preparing them must be sensitive to the teacher/learner dyad, but they must also have a sense for geography as a way of thinking before they can bring the geographical world into the classroom. It is perhaps because of the rather tenuous hold the geographers have had on their profession that they are so quickly experiencing doubts about the world of ''information for its own sake'' that lies ahead.

Certainly, the idea that nothing but good can come from cruising the ever-lengthening ''information highway'' has taken hold in most of our schools, even if the money for this considerable investment is not in sight. Meanwhile, in an increasing number of homes, parents are eager for their children to get a leg up on cyberspace (and perhaps their neighbor's children) and therefore are investing in basic equipment and numerous and expensive programs which they expect will supplement their children's school experience.

As we travel this much-touted highway, teachers and parents concerned particularly with geographic literacy should take note of some general aspects of the computer world before jumping headlong into spending the kind of money accessing cyberspace is going to require. GIS data will doubtless be found increasingly useful in college and university classrooms, regardless of the geographical tradition. High school teachers with academic backgrounds in geography or, at the least, in GIS itself (admittedly a rare breed in any case) will find ways to employ this undeniably rich resource in their teaching. For the vast majority of teachers, however, cyberspace will be restricted largely to materials prepared by others and, probably with some increase in availability, on the Internet, the ever-expanding worldwide web, which touches on every conceivable topic from making available the riches of the vast accumulations of our libraries to the most mundane.

What do we see in this regard? First, for our elementary and secondary schools, much of the on-line material consists of a more or less direct translation of that which is already available in book form. Is it reasonable to ask to what advantage one finds within a CD-ROM holding the contents of an encyclopedia that is not available in its standard printed form? The most likely answer would I think be that *novelty* is the primary attraction, since it is essentially the same thing but in another form; it also makes possible the copying of material directly onto a printer, a dubious advantage over the age-old practice of copying from an encyclopedia or other reference to satisfy homework assignments. Neither appears a particularly attractive reason for such a purpose. There are other novelties that can be built into a CD-ROM program that are not available to the user of a bound reference volume, of course, but the question needs to be asked here, as with much of what is becoming available, how long can we rely on novelty to run the educational ship? We need to realize, as well, that these materials from cyberspace are not infrequently considerably less flexible as sources of information than a good book. It is clear, for example, that compact disks have very specific limitations regarding the amount of information that can be stored on them since so much space is taken up with commands allowing "readers" to move back and forth within the data contained therein—matters with which book readers deal quite naturally as interest lead them from one idea to another. Despite claims that cyberspace will eventually make books obsolete, a growing number of critics are disputing that assertion (e.g., Hiltzik, 1995; Schrage, 1990). In a particularly comprehensive critique of what he calls the "computer delusion," Todd Oppenheimer asserts:

> [T]here is no good evidence that most uses of computers significantly improve teaching and learning, yet school districts are cutting programs—music, art, physical education—that enrich children's lives to make room for this dubious nostrum, and the [present] . . .

administration has embraced the goal of ''computers in every classroom'' with credulous and costly enthusiasm. (Oppenheimer, 1997, p. 45)

While one must grant that interest is critical to learning, both initially and in long-term memory, it has been recognized (literally for centuries), that the kind of interest required for learning to be both meaningful and relatively permanent grows from personal involvement with ideas. Novel situations often attract in the beginning but are essentially ephemeral, with little staying power. However, an almost universal characteristic justifies the new technology on the basis that students find them ''fun'' or otherwise pleasurable in their own right. There can be no argument that when interest is high and the information is relevant, learning becomes a pleasurable experience. But the problem here is that ''fun'' is seen as the initiator rather than the consequence of learning, thus raising the long-debated role of extrinsic forms of motivation over what are generally conceded to be the greater value of intrinsic interest. Many teachers nevertheless rely heavily on extrinsic motivation in teaching, and for those so inclined, computer-generated instructional programs will appear particularly attractive.

Anyone who has not yet discovered how expensive a toll road the information highway is will have a rude awakening once embarked upon it. We are provided an example in a program (*Crossroads USA*, Didatech Software) produced for CD-ROM that purports to teach geography by taking students (grades 4 to 9, plus ''ESL and survival math'') on a make-believe cross-country trip through various parts of the country, during which they play the role of driver of an 18-wheel truck, picking up and delivering various commodities appropriate to a geographic area. The motivational circumstance in this program is a ''game,'' which is ''never predictable and sufficiently complex to hold the interest of players up through high school age,'' assuring us that it is appropriate for a very wide age group. Meanwhile, we are also assured that students will have ''fun,'' emphasizing here the extrinsic values involved in the ''software.'' ''Players'' are led through the program with a series of drawings, charts, and maps (many of which are drawn without reference to their location relative to the ''lower 48''—Hawaii and Alaska are not included). The cost for this particular instructional material is high. Following the general pattern of this new form of educational commerce, its adoption requires the payment of a licensing fee, purchase of compact disks for operation on a restricted number of computers, but with options, as in regular publishing, that provide a lower per-unit cost for quantity purchases. The start-up price for an individual school nevertheless amounts to about $500, sufficient for the purchase of a substantial number of library books. This example is illustrative rather than unusual.

A problem accompanying every decision to purchase within the framework of the new technology is obsolescence. Many of the computers that have so far found their way into the classroom are not capable of handling the programs currently being produced. The continuing and rapid technological advances in the capability of the ''hardware,'' the computer itself, to ''run'' the instructional programs means that there will, for the foreseeable future, be many situations in which programmers (tempted, as one might expect, to create materials that utilize the latest technological developments) will be writing instructional products with limited applicability. The problem of upgrading existing equipment, while relatively small compared to that of equipping the huge number of classrooms waiting to join in the technology revolution, is nonetheless significant and will grow as more classrooms are supplied with computers which are then superseded by later, ''more powerful,'' ones.

Beyond the technical and financial issues lie others that require consideration. I refer here particularly to the tendency of the computer to change learning from a hands-on, socially constructed event to one that is carried out essentially in isolation within a two-, rather than three-dimensional, world. In the first instance, although students may at times work in pairs or even small groups, it is only too obvious that operating a computer is basically a one-on-one situation. Even ''games'' pit one person against all the other ''players.'' In our future, therefore, what will we ultimately see: 25, 30, or more computer ''stations'' making up the typical classroom environment, much as we used to see language laboratories in which the classroom was, literally, made up of a series of ''isolation booth'' cubicles equipped with recording/listening devices? If that scenario is too absurd, then what is reasonable? Should it be an on-demand situation? How might one imagine a program such as *Crossroads USA* functioning in a classroom? Should it be a central activity in the curriculum, a peripheral add-on, or something everyone will ''do'' at one time or another? The operative question here is whether the new technology will simply become an even more efficient means for controlling the curriculum than the traditional textbook or assume some kind of yet undefined ancillary role. There is considerable evidence to suggest that the former option may be the one realized—the computer program may replace the textbook, but in an even more complete way. Because of the expense involved, school personnel will, quite naturally, expect that it be used. But perhaps the more potent factor here is that these curriculum materials will be much more efficient in replacing what it is we expect teachers to be doing in the classroom. Even when the textbook serves as the principle determiner of what is going to be considered in a classroom, much as that form of teaching has been criticized in the past, there is at least a role for the teacher to interact at one level or another with students.

In the second instance, geographic literacy is built on the ability to think within a three-dimensional spatial environment whose characteristics consist of two basic components. One of these derives from the physical facts of location, the other from those relative factors that human occupancy creates. Unfortunately, the computer screen, while also rendering that reality within a basically unrealistic, two-dimensional frame, almost universally deals in representations that themselves are abstractions, as in for example, the drawings of physical realities rather than photographs used in *Crosscountry USA* and the various other kinds of abstractions that can be generated by computers in endless streams.

Indeed, ''real'' pictures are not easily adaptible in computer programming. Experiments in *virtual reality* (the operative word here is the first, *virtual*, meaning not quite real) make it possible to ''see'' constructs of reality, but not the thing itself. In this technology the two-dimensional obstacle is overcome, so the three-dimensional problem is to a certain extent a soluble one. However, virtual reality operationally requires equipment far in advance of anything that will be available in schools for years to come. Even then, it will be available only for an individual user and will require extremely powerful technology. For some uses, particularly architectural designing of various kinds, and increasingly in planning uses for a specific environment, its value is already proving to be great. Nonetheless, like other interactive forms of representation, the images are not those of ''real things'' but are, rather, constructs of reality. Virtual reality has now entered the field of entertainment, and in that aspect we doubtless will see arcade-style auto-racing, down-hill skiing, bungee-jumping, and similar games becoming popular.

Beyond the problem of learning about a three-dimensional world within a two-dimensional format lies an even more critical matter. To reveal the problem it presents, I turn to the fact that the cyberspace world is almost totally dependent on the visual process. Although sound will increasingly become part of cyberspace as technology advances and will be incorporated in newer instructional programs, the primary thrust will continue to be upon visual imagery, and then particularly upon its receptive, or passive, aspect. It is popularly, but incorrectly, thought that the visual process provides the primary avenue for learning. We say we learn primarily through sight, failing to recognize the fundamental importance of hearing to all learning. Without an intact auditory system, normal speech becomes impossible. The profoundly deaf person is, as a consequence, largely isolated from the broader society. Deafness also impedes overall intellectual development because it prevents the development of normal speech, the avenue that people with intact sensory systems utilize to communicate with one another (see, for example, Vernon & Andrews, 1991). There is a tendency to forget that intellectual development generally involves an ongoing interaction, primarily between sight and hearing, but as well through the other senses. Although persons deprived of either hearing or seeing, or both as in the case of Helen Keller and a very small proportion of the population, seem quite naturally to enhance their powers of perception through their remaining sensory pathways, there is no satisfactory substitute for hearing in both socialization and intellectual development. If one had a choice between whether one might be without sight or without hearing, there can be no doubt that being sightless is much the lesser burden, much as we might at first think otherwise, for a blind person can, for all practical purposes, do what a sighted person is capable of, from playing a musical instrument to typing on a keyboard. Just because one has an intact sensory system does not mean that hearing as a consequence plays only a minor role in thinking and learning, although one might be hard-pressed to defend much of the typical school curriculum where sight and silence are so often emphasized.

As we learn to avoid the caveats of Postman's *Technopoly*, there remain important uses for the new technology where geographic literacy is concerned. It would seem self-evident that the Internet, the worldwide web, and the many other sources of information of various kinds that can be accessed "on-line" provide a rich source, perhaps not so much for the younger student, but surely for the teacher. Increasingly, students will come to school from "computer-literate" homes. The first responsibility of the teacher, it seems to me, will be to achieve some degree of that form of literacy, ahead of the student clientele if possible, but certainly with it. I do not believe that the computer itself, as a piece of equipment in the classroom, is going to be the direct resource that many think. It will, of course, prove to be a valuable source of information, first to the teacher, and second, to those students who are capable of abstract thinking. As long as it is viewed primarily as a substitute (albeit a potentially more interesting one) for a textbook or standard reference or as a game, there are good reasons to opt instead for what we might call an interactive classroom, meaning one in which the teacher and students work on problems and issues on a first-hand basis and where *cyburbia* (as the Internet is often called) is utilized as a resource for learning rather than an end in itself.

## SOME THOUGHTS ON THE FUTURE OF GEOGRAPHIC LITERACY

I have throughout this discussion expressed my concern that advocates of instruction in the geographic strand of the school curriculum, by insisting that it must have its own identity as a separate subject taught in rather magnificent isolation (an autonomous "subject" valid in its own right), may not be following the most fruitful path toward the goal of increasing geographic literacy. I have pointed to the fact that there are almost an infinite number of opportunities throughout the curriculum where matters geographic might be given honest attention. Saying this, I am also very aware that curriculum development at the turn of the century—how particular subjects find their niche within the curriculum itself—is quite a different matter than it was a generation or so ago. Then there was serious discussion about how to create a balanced curriculum, meaning one that reflected an appropriate weighting of all the things society seemed to expect of its most important social institution. Today the story is not one of achieving a balance between the various goals that the school has traditionally sought to reach. It is, rather, one of competing interest groups seeking to find their particular place in the curriculum sun.

The result is a badly skewed curriculum in which instruction in skills has largely overtaken the idea content of the curriculum and in which the problem of print and computational literacy has virtually banished ideas in their own sake from the curriculum of the elementary school. We find its parallel in the secondary school where students are being segregated according to achievement. This is not simply being done between classes; now, with the rapidly rising number of specialized high schools, the intellectual elites are syphoned off for preprofessional training or advanced preparation for college, leaving the majority to a curriculum almost exclusively devoted to a continuation of the skills development emphasis of the elementary years.

It is within this milieu that efforts to improve literacy levels where geography is concerned must function. Given the tradition (now exacerbated by the trends I have noted above) of a top-down approach to curriculum development, it is not surprising that much of the national effort to bring about improvements in geographic literacy are based on this principle. National organizations—in this instance, primarily the National Geographic Society—provide seed money. This leads to the formation of "alliances" at the state level, which then turn their attention to grass-roots activities of various kinds. Top-down curriculum development also usually includes the preparation of instructional materials designed for individual teacher use.

This approach, while widely practiced for a long time in American education, is most notable for its lack of success. I believe this has been the case primarily because top-down practice is essentially an attempt to bypass the teacher in the teaching/learning dyad. Instruction thus is viewed primarily as an interaction between student and the textbook, or, more recently, other avenues for learning: workbooks, computer programs, and so forth. There has been a long, evolving tradition in American education of relying heavily on the material side of the instructional equation in the educative process. Geography texts of the nineteenth century provided the questions with their appropriate answers, and teachers were expected to conduct memoriter-based daily lessons based on the textbook. Since that time there has been a steady trend toward expansion of the notion that teachers should be given detailed instructions regarding the best ways to utilize the textbook or other instructional material to best advantage. The idea that the teacher is best left to his or her own devices is foreign to many school administrators. Perhaps the best example of directing the curriculum traffic is found in the "teacher's manual" that accompanies every "basal reading series," but several recent publications following similar patterns have appeared that are intended to substitute for the teacher's lack of knowledge about geography and how geographic study can be carried out. See, for example, Ludwig et al., *Directions in Geography* (1991); Stansfield, *Building Geographic Literacy* (1992); Slater, *Learning through Geography* (1993); Gersmehl, *The Language of Maps* (1991).

Whether it is possible to make up for the lack of a sense for geography in this fashion is certainly open to question. *Teachers teach what they know*, and so it would seem probable that these are not much better than stopgap measures. The larger question is how to sensitize a profession to the need for balance in the curriculum and to an understanding of the role of geographic inquiry within that total picture. Many years ago (1934), Lucy Sprague Mitchell wrote an illustrated pamphlet for the New York City schools titled, *Young Geographers: How They Explore the World and How They Map the World*. In her introduction she wrote:

> This discussion of Young Geographers is divided into two parts. The first part maintains the position that even young children can and do think in geographic terms, and suggests how schools can plan to give real opportunities to their young geographers. Both the general thesis and the school programs and procedures are the result of many years of working with geographers, ranging from two through thirteen years. After this age, the laboratory approach in geography is generally recognized as desirable though, even then, too seldom followed. The second part of the discussion deals with the map-thinking and map-making aspect of the work of these same young geographers and their teachers. (p. 7)

Whether one uses the term ''laboratory approach'' or ''holistic curriculum,'' what is implied is a classroom in which the teacher sets broad outlines of what is to be accomplished but then adopts a teaching style that encourages the students to pursue knowledge within their own frame of reference. Many years after Mitchell, Janet Kierstead (1984) analyzed four classrooms that met criteria of outstanding effectiveness. Three criteria were applied: (1) nomination by outside professional sources; (2) nomination by the building principal; and (3) determination of the effectiveness of the nominated classrooms, judged by their superiority of test scores. Two major findings emerged from her study. First, these particular teachers held without exception or qualification to the belief that all children are interested in and capable of learning. This may seem rather unremarkable; surely all teachers hold to this view. Unfortunately, however, that is not the case. Not a few are not only pessimistic about, not only individual children, but their entire class. Holding a positive view of human nature is, to some, a difficult proposition, particularly given the problems of many inner city and even suburban classrooms. Second, the teachers in this study had established systems of classroom management and organization

> in which the teachers attended to students according to need and allowed them to share in responsibility and control over learning. That system was comprised of four components: sequential processes for developing skills, strategies for ensuring student accountability, strategies for monitoring and guiding student growth, and a supportive environment of resources. (p. iv)

Critics of the holistic or laboratory classroom are quick to complain that descriptions such as this suggest a fulsome, chaotic, out-of-order environment, in which students engage in any kind of activity but the ones intended. Others recognize the validity of such a classroom but are unable, at least at the moment, to realize them much as they might wish to. Others will argue that this may be a feasible way of functioning with young children (both Kierstead and Mitchell are describing classrooms of young children), but hopelessly inappropriate for learning when things get serious (i.e., after the elementary school years). All are arguments based on a philosophy of human behavior, the teacher's belief about the ability and conditions under which learning best occurs. No minds are going to be changed with mere argument. Both Mitchell and Kierstead, like their practitioner counterparts in numerous classrooms today, are primarily concerned with the quality of learning as we conventionally think of it, but they have a different view of how quality can be achieved.

I believe there is a great deal of validity in approaching the problem of geographic literacy within this kind of a broad framework. After all, geography is more a way of thinking than a particular body of subject matter, although there are indeed specific tools that accompany and in fact make possible higher levels of thinking where spatial interactions are concerned. Geographic educators have been a long-suffering group, however, and when the opportunity presents itself (as is presently the case with the financial backing of organizations such as the National Geographic Society) and combines with a public holding the opinion that "geography" (usually defined as an ability to answer the Mozambique Question but with a much broader potential for stimulating intellectual inquiry), it is not surprising to see advocates for "bringing geography back" going for the jugular (i.e., staking a claim for a piece of the curriculum devoted solely to their "subject"). We have of course been approaching educational problems in just this fashion for quite some time, with not even minimal results. As we have focused more sharply on skill development in print reading, for example, by lengthening the instructional time devoted to it and teaching it in ever greater isolation from the rest of the curriculum, reading scores have failed to respond according to expectations, much to the surprise of those who fail to understand the organismic whole of the total curriculum—that what happens in one aspect will have an effect elsewhere if the problem of teaching is approached with the understanding that the whole is equal to more than a sum of its parts. Because geography is an integrating discipline, we would be well advised, I think, to approach its teaching within the broad frame that this fact implies.

> It is in fact nothing short of a miracle that the modern methods of instruction have not yet entirely strangled the holy curiosity of inquiry; for this delicate little plant, aside from stimulation, stands mainly in need of freedom; without this it goes to wrack and ruin without fail.
>
> I have no doubt but that our thinking goes on for the most part without use of signs [words] and beyond that to a considerable degree unconsciously.
>
> Albert Einstein

## REFERENCES

Cuban, Larry. 1993. *How teachers taught: Constancy and change in American classrooms, 1890–1990.* New York: Longman.

Donaldson, Margaret C. 1978. *Children's minds*. New York: Norton.

Downs, Roger M., & Liben, Lynn S. June 1991. The development of expertise in geography: A cognitive-developmental approach to geographic education. *Annals of the Association of American Geographers*, 80, 304–327.

Gersmehl, Philip J. 1991. *The language of maps*. Indiana, PA: National Council for Geographic Education and Indiana University of Pennsylvania.

Hanna, Paul R. 1987. *Assuring quality for the social studies in our schools*. Stanford, CA: Hoover Institution Press.

Hiltzik, Michael A. December 13, 1995. Lesson no. 1 on educational software: It does not work. *Los Angeles Times*.

Husen, Torsten, Ed. 1967. *International study of achievement in mathematics: A comparison of twelve countries*. New York: Wiley.

Kierstead, Janet. 1984. *Outstanding effective classrooms: A study of the interdependence of compositional, psychological, behavioral, and organizational properties in four primary classrooms*. Unpublished doctoral dissertation, Claremont Graduate School, Claremont, CA.

Ludwig, Gail S. et al. 1991. *Directions in geography: A guide for teachers*. Washington, DC: National Geographic Society.

Mitchell, Lucy Sprague. 1934. *Young geographers*. New York: John Day Company.

Oppenheimer, Todd. July 1997. The computer delusion. *Atlantic Monthly*, pp. 45–56.

Postman, Neil. 1982. *The disappearance of childhood*. New York: Dell.

Postman, Neil. 1982. *Technology: The surrender of culture to technology*. New York: Random House/Vintage Books.

Ravitch, Diane. 1989. The plight of history in American schools. Chapter 3 in Paul Gagnon & the Bradley Commission in History in the Schools, eds., *Historical literacy: The case for history in American education*. New York: Macmillan.

Saxe, David Warren. 1991. *Social studies in schools: A history of the early years*. Albany: State University of New York Press.

Schrage, Michael. October 8, 1990. The networks of educational hell are wired good intentions. *Los Angeles Times*.

Slater, Frances. 1993. *Learning through geography*. Indiana, PA: National Council for the Social Studies.

Stansfield, Charles A., Jr. 1992. *Building geographic literacy: An interactive approach*. New York: Macmillan.

Sui, Daniel Z. November/December 1995. A pedagogic framework to link GIS to the intellectual core of geography. *Journal of Geography*, 94.6, 578–591.

Suransky, V. P. 1982. *The erosion of childhood*. Chicago: University of Chicago Press.

Vernon, McCay, & Andrews, Jean F. 1990. *The psychology of deafness: Understanding deaf and hard-of-hearing people*. New York: Longman.

Virginia State Board of Education. 1934. *Tentative course of study for Virginia elementary schools. Vol. 1, Grades I–VIII. Vol. 2, Grades IX–XII*. Richmond, VA: Division of Purchase and Printing.

# Afterword

Not so long ago, there was concensus among professional educationists that the only serious impediment to school improvement was a lack of interest and attention. Not only was it possible to bring about such change; in a sense, change was seen as inevitable. Enlist the recalcitrant, and there would be smooth sailing ahead. This was the Progressive view, which was inherited from the great movers and shakers of the post–Civil War period (John Dewey, William James, G. Stanley Hall), the birthing period of schools as we know them today. However, a funny thing happened on the way to that forum of change. From the robust experimentation that characterized the first half of the twentieth century, a destructive cynicism gradually arose. Fueled by two real wars, World War II and the Korean War, exacerbated by the perception that an inevitable war was yet to be waged (the war of the Red Scare), and then further damaged by a conflict that, in short order, divided the nation (in Vietnam), the previously widely held belief that change and progress were easily comingled concepts where the schools were concerned began to dissipate.

It was at this juncture that the public view of what constituted an appropriate education began to polarize and an oscillation began between those points of view, one rooted in the earlier developmental agenda of the Progressives, the other a romanticized reinterpretation of a past that had never really existed, in which everyone learned whatever was set before them. Thus began an ongoing disagreement between those who would advocate a "return to the basics," as if that were a reality that enjoyed a commonality of understanding regarding its meaning, and those who described themselves as "constructivists," meaning those who placed the process of learning ahead of, or at least equal to, the matter of knowing (or, as those siding with what might also be called the traditionalists would put it, knowledge of subject matter).

Of course much of this was, in large measure, theoretical talk. Schools are, by definition, as hidebound in their ways as any social institution could possibly be and as a consequence are highly resistive to change. Moreover, although some would have it to the contrary, schools have remained conservative in their orientation and more devoted to "the basics" than many would believe. Nonetheless, since the two real wars, some broad directions have emerged over the years that have influenced public policy where the American educational enterprise is concerned. And it is within this context that the geographic strand of the school curriculum must find a way to flourish.

One of the most potent influences of which the geographic educator should be aware is the growing reliance on standard tests as measures of student competency and, as a consequence, teacher effectiveness. The movement toward setting educational goals according to the results of standardized tests is, in many respects, no more than twenty years old, having grown in importance only since the last period of major curriculum experimentation, during the late 1960s and early 1970s. Nonetheless, there is now a widespread acceptance among legislators and other public figures that the data derived from standardized tests, especially those administered to very large (including statewide, if not nationwide), populations, provides the only valid measure of the success of the American school system.

However, the problem for the geographic educator lies not so much in the invidious comparison—after all, there are no tests of "geography" to show trends or other kinds of comparative data. It lies in the fact that the tests we do have are concerned with matters other than the social education of students; and these are, primarily, in something called "reading," and math (meaning, for the most part, computational skills, along with the beginnings of something we might truly call mathematics). Just as power follows money, then, we have a situation where other matters are taking precedence, even as we bemoan the lack of instruction where geography is concerned.

The acceptance of testing as an appropriate device for assessing the effectiveness of our schools means, therefore, that there will be an increasing threat toward destabilizing the curriculum and, in fact, of disrupting the educative process to the extent that, as educational goals become synonymous with the content of tests, any view of learning as a creative act will be lost.

The first reaction to any seeming effort to downgrade or even eliminate one's pet area of the curriculum often seems to be to adopt a bunker mentality: in school parlance, I make room for my "subject" no matter what it may cost others with a different curriculum agenda. I have argued that this is not just unrealistic where the geographic curriculum is concerned, but futile in view of the growing recognition that the old subject matter boundaries are eroding with the onrush of advances in knowledge about our universe. Knowledge is a holistic, not a particularistic, matter.

I have asserted that teachers teach what they know. Thus, perhaps the most persistent dilemma facing anyone concerned with the problem of geographic literacy is how the necessary knowledge and skills in teaching can be made available to the practitioner in the classroom. Traditionally, we have chosen to short-circuit the problem; that is, we have tried to supplant teacher knowledge with various tools and accessories that, it is mistakenly thought, provide a legitimate substitute for the real thing, namely, the teacher's familiarity with what he or she was expected to teach. The basic tool in this regard has been the textbook. Originally, as we have seen, textbooks were objects of memorization, but as time passed and it became recognized that rote learning was less than effective, writers of textbooks began inserting directions and suggestions to the teacher. Then, with advances in printing technology, teachers were showered with "worksheets" and other "supplementary" materials, a process that continues today. With the invention of the mimeograph and hectograph, the ability to copy previously printed document marked the opening of a paper onslaught that students increasingly bear today. Now, of course, laser printers, scanners, and all sorts of other devices make it possible to reproduce previously published material (although copyright laws have been invoked to curb their pirating, this stringency is probably observed more often in the breach than in actuality). But the point remains: these kinds of materials are intended to compensate for what is usually perceived to be the teacher's lack of knowledge or skill in teaching.

Some of these teacher substitutes are perilously similar to the kinds of irrelevant things that have been circulating around geography curriculums for many years. As long as school personnel rely on using instructional materials created far from the classroom by people whom no one knows and who know little to nothing about the needs and interests of students in a particular classroom, I doubt that we can expect much improvement in geographic learnings since, for one thing, they tend to perpetuate traditional but narrow and unacceptable views of the nature of geography. There is, in other words, no satisfactory substitute for a teacher's knowledge and love for geographic inquiry. Nonetheless, there is a widespread and apparently growing acceptance of the belief that developments in the media world will mitigate any prior deficiencies in the materials available to "help" teachers teach.

This book stands as one attempt to introduce to those who teach a way of thinking about the world in spatial terms, the essence of geographic thinking. What vehicle, or subject, one elects to use for this purpose is, unimportant other than to say there must be something substantive to undergird the process and a rationale for its inclusion in the curriculum. The teacher's experiential background is important, but so are the circumstances of students' lives: where they live, what experiences they have in common, and how developed are their thought processes where spatial ideas are concerned. These are matters that cannot be anticipated by textbook writers or others preparing teacher materials.

There is a growing list of resources available to help teachers understand the nature of geographic inquiry, some fraught with the problem caused by offering materials that will substitute for a teacher's lack of knowledge and understanding of geographic phenomena (i.e., the teacher's own level of geographic literacy).

One potential source that has the opportunity to avoid the quick fix is the system of Geographic Alliances (referred to in Chapter 1). But the problem of having too few knowledgeable professionals to the number of classroom teachers continues.

In bringing together several seemingly disparate areas of inquiry—psychology, history, education, geography—I have attempted a fusion that, as an educationist, seem to me to be absolutely essential to productive planning for the geographic strand of the school curriculum. And although I have argued throughout that traditional academic boundary lines such as the four I mention here are losing their distinctiveness, there is always the possibility—even the probability, at least in some quarters—that turf watchers will take issue with some, or even many, of the thoughts on the subject of geographic literacy I have presented.

It is quite apparent, nonetheless, that the American school system has embarked on what appears to be a chancy journey in which the idea of inculcating skill precedes all else. That *form* follows *function* is, once again, a maxim too often observed in the breach. Obsession with skill development (i.e., the acquisition of specific bits of information) results in a kind of mental disconnectness that has a very short shelf-life. The foundation of geographic literacy will not be built on the ''Where is its?'' and ''What is its?'' of the world. Rather, it will grow from personal experiencing of the environment. From that kind of involvement, the *forms* of geography will become meaningful tools to furthering individual competence.

# Index

**About the Author**

MALCOLM P. DOUGLASS is Emeritus Professor of Education, Claremont Graduate University, Claremont, California. His previous books include *Social Studies: From Theory to Practice in Elementary Education*, *Reading in Education: A Broader View*, and *Learning to Read: The Quest for Meaning*.

ISBN 0-275-96138-9
EAN
9 780275 961381 90000>
HARDCOVER BAR CODE

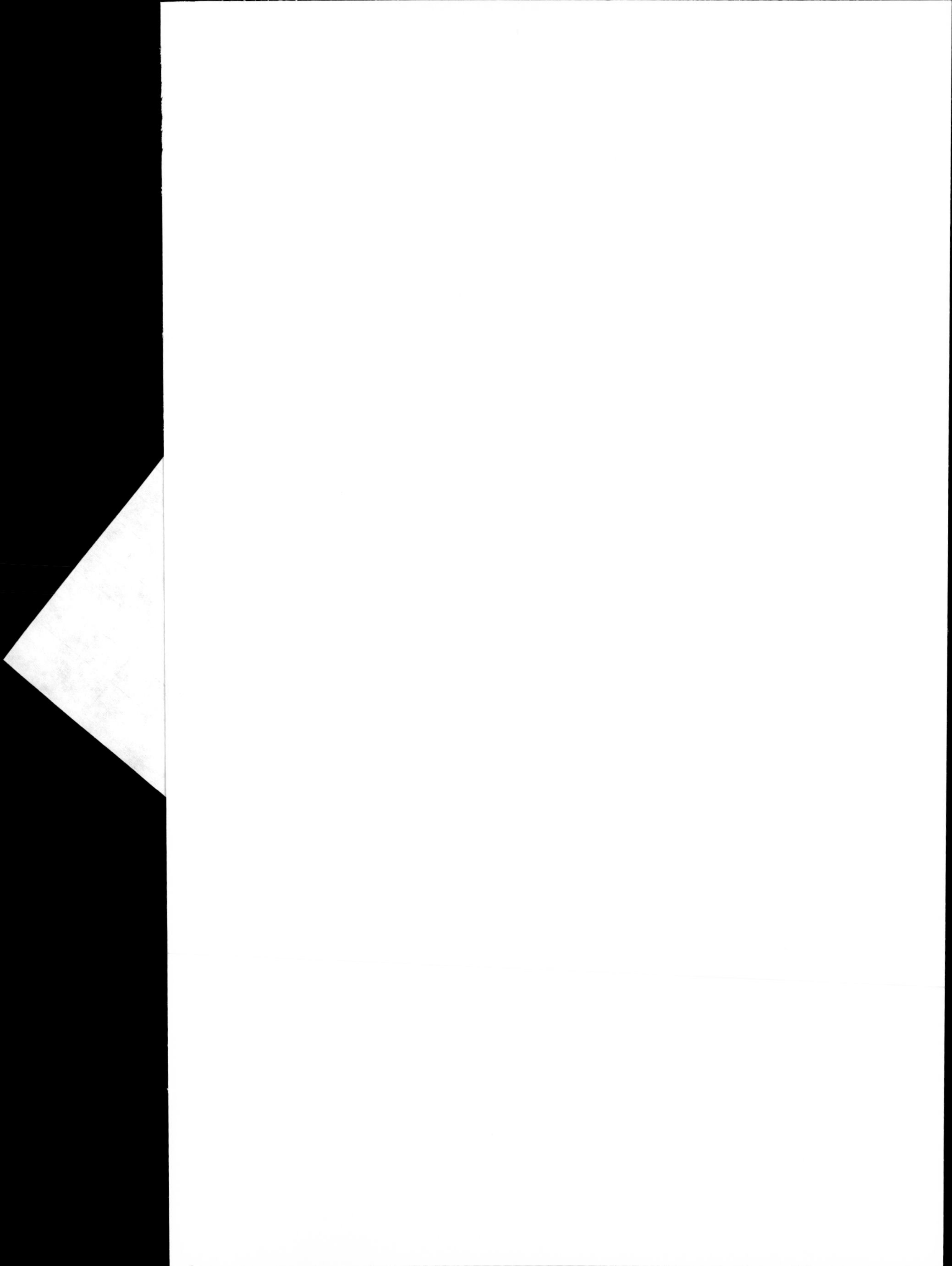

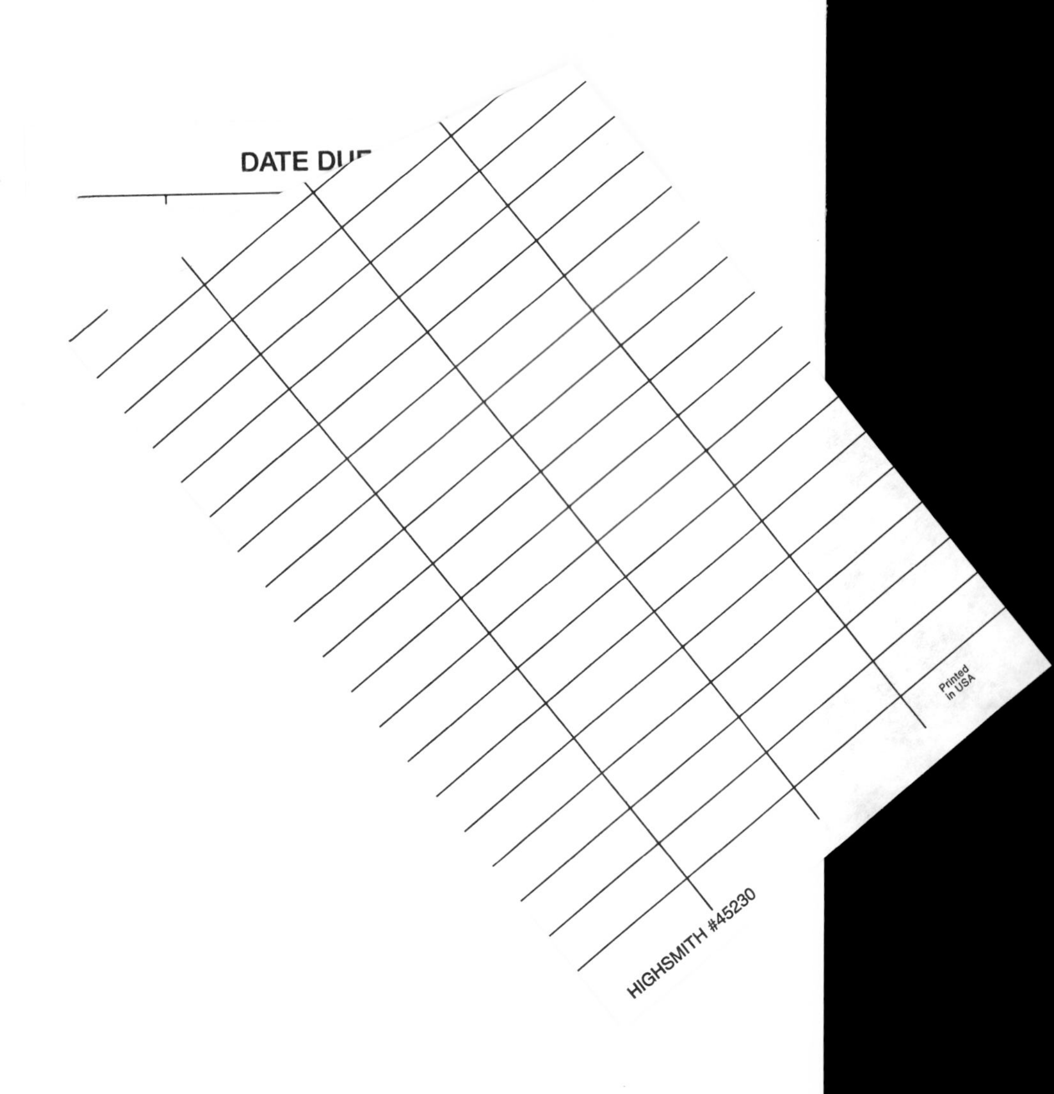
DATE DUE
HIGHSMITH #45230
Printed in USA